A Random Walk in Physics

Massimo Cencini · Andrea Puglisi · Davide Vergni ·
Angelo Vulpiani

A Random Walk in Physics

Beyond Black Holes and Time-Travels

 Springer

Massimo Cencini
Istituto Sistemi Complessi
Consiglio Nazionale delle Ricerche
Roma, Italy

Davide Vergni
Istituto per le Applicazioni del Calcolo
Consiglio Nazionale delle Ricerche
Roma, Italy

Andrea Puglisi
Dipartimento di Fisica
Università di Roma Sapienza
Roma, Italy

Istituto Sistemi Complessi
Consiglio Nazionale delle Ricerche
Roma, Italy

Angelo Vulpiani
Dipartimento di Fisica
Università di Roma Sapienza
Roma, Italy

ISBN 978-3-030-72533-4   ISBN 978-3-030-72531-0   (eBook)
https://doi.org/10.1007/978-3-030-72531-0

This Springer imprint is published by the registered company Springer Nature Switzerland AG
The registered company address is: Gewerbestrasse 11, 6330 Cham, Switzerland

*To all who were not afraid to stay out
of the mainstream: Ludwig Boltzmann,
Cheikh Anta Diop, Antonio Gramsci,
Lev Davidovich Landau, Rosa Luxemburg,
Lewis F. Richardson, and the others.*

Acknowledgments

The four authors of this book have worked together for decades, as part of a larger scientific enterprise of theoretical physicists—perhaps more a brotherhood than a research team—that bears the name of TNT group, based at Sapienza University in Rome. They owe a debt of gratitude to all friends, colleagues, and collaborators of TNT group for developing the vision which underlies the writing of this book. In particular, they wish to thank Fabio Cecconi and Massimo Falcioni, who participated in the early stage of this book and with whom they had long discussions on how to popularize scientific ideas which are often disregarded by the general public. The authors also acknowledge the invaluable help of friends such as Marco Baldovin, Francesco Borra, Claudio Castellano, Andrea Plati, and Marco Zannetti for reading parts of the book and suggesting improvements. They are also grateful to Gianni Battimelli, Luca Biferale, Roberto Natalini, Michele Buzzicotti, the Archivio Amaldi, and the Volterra Family Archive for providing us with some of the figures.

Contents

Chapter 1
A Sort of Introduction

If we study the history of science we see produced two phenomena which are, so to speak, each the inverse of the other. Sometimes it is simplicity which is hidden under what is apparently complex; sometimes, on the contrary, it is simplicity which is apparent, and which conceals extremely complex realities. (Henri Poincaré)

To develop the skill of correct thinking is in the first place to learn what you have to disregard. In order to go on, you have to know what to leave out: this is the essence of effective thinking. (Kurt Gödel)

The first duty of the authors of a book is to explain their reasons, which means answering the question *why this book?* In order to convince the reader that there are some good motivations, we start from an old story: Friedrich Wilhelm IV (king of Prussia in the period 1840–1861), owing to his enthusiasm about science, was often asking his royal astronomer *"Herr Argelander, anything new happening in the sky?"* Once the astronomer, having no new interesting result to offer to the king, bravely replied *"Does Your Majesty already know the old things?"* We have the feeling that after two centuries the situation did not change too much. Usually, people far from the academic world do not know (or, reading the scientific breaking news, only have a vague often misleading idea about) the latest scientific discoveries: well, in our opinion, this is not a serious problem. Unfortunately, they ignore also some basic results of older science as well as the ideas underlying applications they use in their everyday life. Moreover, often non-scientists have a foggy or stereotypical idea of what science and research are. A possible reason is the common belief of many authors of popular scientific books that all readers, as the King of Prussia, are avid only of the last discoveries, or of the most spectacular and exotic[1] ones. Therefore, as a net result, we have a plethora of popular books covering topics like black holes, time-travels, gravitational waves, big bang, Higgs boson, and multi-universes.

---

[1] Unfortunately, often the focus on exotic and far from everyday life science contributes to the common belief that most scientists conduct strange researches very far from practical (and economical) needs.

© Springer Nature Switzerland AG 2021
M. Cencini et al., *A Random Walk in Physics*,
https://doi.org/10.1007/978-3-030-72531-0_1

This book does not follow the mainstream: its purpose is to give an informal and easy account of some ideas of modern physics and mathematics that, in spite of their major conceptual and practical relevance, are not typically discussed in popular texts. Moreover, we present also a brief account of the life and personality of the scientists who contributed to the development of those ideas. The potential readers we had in mind are high school teachers, college and high school students, and—more generally—persons with just a basic knowledge of mathematics (high school), but with interest and curiosity in science. Our ambition is to provide the readers with some friendly tools to better understand the world around us as well as to get some history of the ideas that changed it. In our path we tried to avoid too technical details (only in a few cases we introduce some simple formulas or more technical footnotes), without the abdication to correctness of the conceptual aspects. We will mainly focus on a set of topics covering the following subjects (and their interplay): statistical mechanics, soft matter, probability, chaos, complexity, and the role of models and numerical simulations. The various topics have been selected for their conceptual and practical relevance and, of course, within our areas of interest and expertise.

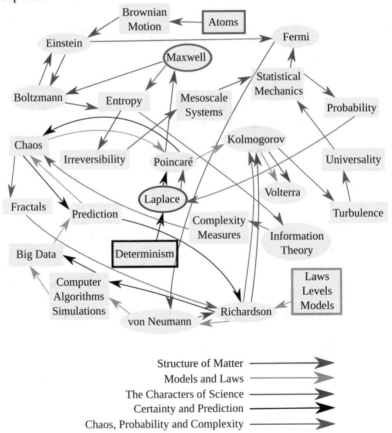

The book consists of 28 entries, pictorially depicted in the above graph. Some of the entries are devoted to *specific arguments* (the rectangles), others to several *important scientists* (the ellipses), including those who, in spite of the importance of their contributions to science and the history of thought, are almost unknown to the general public. Unlike artists and writers, with very few exceptions (A. Einstein, S. Hawking and very few others), the life of scientists is not well known among the non-experts. We do not understand why: many scientists had an intense and interesting life, sometimes with vivid (occasionally tough) contrasts with their competitors. As a paradigmatic example we can mention Ludwig Boltzmann and his battle for the atomistic hypothesis against Ernst Mach and the energetic school. A short presentation of the main aspects of life of some important protagonists of science can also be useful to understand and appreciate their contributions and the way science proceeds in practice.

Deliberately the entries appear in alphabetic order. With this choice there is no attempt to do something similar to an encyclopedia. On the contrary, our intention is to avoid an exceedingly systematic and potentially boring approach: we invite the reader to perform a *random walk* among the entries, jumping from one entry to another at will. Even if it is not required to follow a specific order in the reading of the book parts, in the figure above we suggest some possible reading paths, linking a few entries according to a common theme or *fil rouge*. As clear from the graph, the itineraries are not disconnected and some entries are common in different paths, in addition their precise order in the sequences is not particularly relevant. We invite the readers to create their own road map of the book. Below we offer a short guide to the above-suggested routes.

Structure of Matter

Atoms → Brownian Motion → Einstein → Boltzmann → Entropy →
Irreversibility → Mesoscale Systems → Statistical Mechanics → Fermi.

Within our daily life, we experience matter has a continuum: apparently it seems that we can divide an iron bar of 200 g, in two bars of 100 g, and then in four bars of 50 g and so on. Today we all know that such an experience is illusory, as matter has a granular, discrete structure because of the existence of atoms. The atomic hypothesis started with a philosophical intuition of Leucippus and Democritus (fifth century BC). However, we have been able to play with atoms and sub-atomic particles only since the first decades of twentieth century AD. The path from Leucippus and Democritus to modern particle accelerators has been tortuous and not straightforward both for some general philosophic (as well as religious) reasons and for the difficulties inherent to the visualization of atoms. At the end of the nineteenth century even eminent scientists, e.g., Planck and Mach, did not believe in the physical existence of atoms. Only after the works of Einstein and Perrin on the Brownian Motion in the first decades of 1900—explaining the erratic behavior of colloidal particles, whose dimension are small (compared with macroscopic objects) but much larger than the molecular sizes—the physical existence of atoms had been accepted by the whole scientific community. The Brownian Motion had an important role not only in the history of the modern physics, as it constituted the starting point of an interesting chapter of probability theory (stochastic processes) with applications in a wide range of fields from biology to finance. In addition, colloidal particles had been the prototype of the so-called mesoscopic systems.

Models and Laws
Laws, Levels, and Models $\rightarrow$ Richardson $\rightarrow$ von Neumann $\rightarrow$ Computer, Algorithms, Simulations $\rightarrow$ Big Data $\rightarrow$ Prediction $\rightarrow$ Chaos $\rightarrow$ Poincaré$\rightarrow$ Komogorov $\rightarrow$ Volterra.

In science one can find terms as laws, theories and models; for instance the law of gravitation, quantum theory, and Lotka-Volterra model. Someone could think that models are less noble or less important than theories, but they are actually unavoidable in scientific practice. We can say that even wonderful theories as Newton's mechanics, Maxwell's electromagnetism, and Quantum Mechanics are actually very accurate and general models. They are able to describe reality only at a certain scale and resolution. Many interesting systems have a multiscale structure, involving very different characteristic temporal and spatial scales, important examples are climate dynamics and proteins. Due to the difficulties in dealing with the variables involved in multiscale systems, the unique way to handle such phenomena is to find a description at a certain level of resolution. On the other hand, it is usually rather difficult to find a methodology to build effective equations and there are no systematic (general purpose) recipes to achieve such a goal. Therefore, it is always necessary to use the previous understanding of the considered system and, often, analogy and intuition. A perfect example of this procedure is given by the history of the weather forecasting, starting from the first empirical approaches, based on weather maps, then to the ingenious ideas of Lewis Fry Richardson to finish with the Meteorological Project at the Institute for Advanced Study (Princeton) led by John von Neumann, Charney, and colleagues in the 1940s–1950s, which gave birth to the numerical weather forecasting today in use.

The characters of Science
Laplace $\rightarrow$ Poincaré $\rightarrow$ Maxwell $\rightarrow$ Boltzmann $\rightarrow$ Einstein $\rightarrow$ Fermi $\rightarrow$ von Neumann $\rightarrow$ Richardson $\rightarrow$ Kolmogorov $\rightarrow$ Volterra.

Science has many faces and the researches on the specific topics can have different origins, from pure curiosity to understand very general aspects of the Nature, to very practical applications. In a similar way, the scientists can have rather different interests, attitudes, and approaches. Remarkably, some topics originally motivated by very abstract problems, after some decades (or even centuries) had been relevant even for very practical applications. Here a short list. Boltzmann's equation, originally introduced to understand thermodynamic irreversibility, is now used for projecting electronic devices and space capsules. The methods introduced by Poincaré for the study of chaotic systems, today are used for the planning of missions in space. Quantum mechanics, which was born to explain black body radiation and certain behaviors of matter at atomic scale, today has a major role in many aspects of our life. Fractals had been introduced for rather sophisticated mathematical problems, now are used even by the movie industry. Looking at the history of science it is not difficult to find several eminent scientists who had an important role even in the political life of their time. Everybody knows the story (legend?) of Archimedes in the war between Syracuse and the Roman army. In more recent times several scientists had been deeply involved in military projects, as John von Neumann, while others, as Vito Volterra and Lewis Fry Richardson, used their prestige and intelligence to contrast the power and the politicians of their times. Pierre-Simon Laplace, for a short period was minister under Napoleon, and Henri Poincaré was involved in the famous Dreyfus affaire. John von Neumann had an important role in the Manhattan project for the first nuclear bomb as well as in other important military activities, while Richardson was a coherent pacifist.

Certainty and Prediction
Determinism → Laplace → Poincaré → Chaos → Prediction → Richardson → Computer, Algorithms and Simulations → Big Data.

Usually science is considered the spring of certainty. The prototypical example begin astronomy, in fact we use to say "astronomical precision" as a synonym of certainty. In the twentieth century the discovery of the planet Neptune, using the laws of motion and gravitation theory, had been seen as the triumph of the ideas of Laplace on determinism and the power of Newton's mechanics. After the discovery of chaos by Poincaré (ironically occurred during the investigation of an astronomical problem), we know that also "astronomical precision" cannot be considered as exact. Even in a deterministic system, the evolution can be chaotic, meaning that infinitesimally small perturbations in the initial state, for instance, a slight change in one body's initial position might lead to dramatic differences in the later states of the system. Such a result at a first glance can sound rather negative, somehow a limit of science, on the other hand, the understanding of the specific aspects of a given chaotic system allows us to focus only on the problems which can have a serious answer. We can say that the real power of science is in understanding the limits of the theories and models, as explained very clearly by Confucius: *"To know that you know when you do know, and know that you do not know when you do not know: that is knowledge"*.

Chaos, Probability and Complexity
Maxwell → Entropy → Information Theory → Complexity measures → Chaos → Fractals → Richardson → Kolmogorov → Turbulence → Universality → Statistical Mechanics → Probability → Laplace

Terms as certainty and probability seem to belong to different realms; in a similar way determinism and complexity appear to be in opposite positions. Surely in all the situations where a large number of causes are involved (e.g., in a gas) we expect a certain degrees of "complexity", and the use of probabilities appears natural. On the other hand, with the discovery of chaos, we understood that even in deterministic chaotic systems, with only a few number of variables, one can have high levels of complexity in which statistical approaches are mandatory. On the other hand, even in the realm of probabilities, one can find something rather similar to deterministic behaviors: for instance, the true essence of the limit theorems (such as the law of the large numbers) is the fact that in presence of many independent variables (causes) one can have a result which is practically sure—in the appropriate limits with probability arbitrarily close to one, in the jargon of probabilists. Such ideas date back to Jakob Bernoulli, at the end of the eighteenth century, with his motto *"Something is morally certain if its probability is so close to certainty that shortfall is imperceptible"*. This result gives a sensible connection between probability and empirical world, stressing the non-abstract character of probability.

Besides the 28 entries in alphabetic order, the reader can find three extra entries. At the beginning a presentation of the Random Walk, a topic which has a relevant role in many fields of modern science and appears in the title of the book. At the end a *divertissement* where a few popular fiction works (books and movies) are put in contact with the 28 entries, suggesting that many concepts and scientists discussed here—even if not frequently present in popular science books—have emerged from the ivory towers of science and already reached a wider audience. Finally, at the very end, we provide the readers with some further readings (technical and non-technical articles and books) organized along the suggested paths, which can be used to deepen some topics of interest.

Chapter 2
Random Walk

In this book we have collected a group of scientists and a set of scientific topics most of which are outside the mainstream of popular science. We have chosen those arguments since they are fundamental elements of knowledge for a large part of modern science even though they are not very much popularized. In this context, the random walk (RW), that gives the title to the book, is a classical example of a scientific topic with very important applications in many different fields from biology to chemistry, from economics to sociology, and we deem it appropriate to start this journey briefly presenting this subject.

The RW, sometimes called drunkard's walk, is a simple random process in which the subject of the action, the walker or the drunk, at each time interval chooses a direction at random, for example, by flipping a coin, and take a step in that direction. Step by step the walker generates a random path. The simplest instance of RW considers a walker that can only move on a line by making steps of fixed amplitude randomly choosing to go to the right or to the left. In this case, assuming that the walker starts from an initial position, by using elementary combinatorics (essentially the binomial coefficients) it is possible to calculate the probability of finding the walker at a certain time in a certain position relative to where it started (Fig. 2.1).

This simple exercise originates a number of interesting consequences. For instance, at large values of times, starting with many walkers in the same initial position, one could wonder how many walkers have reached a given distance from the initial position. Intuitively, one can expect that many walkers remain near the initial position, but the exact answer to this problem is given by the Gaussian probability distribution that is able to tell in which positions walkers are most likely to be found (see the entry Probability). Figure 2.1 illustrates these features. Moreover, considering very small time and jump intervals (in a given and fixed proportion) one obtains the diffusion equation, namely, one has that the average square distance covered by the walker is proportional to the time lapsed (see the entry Brownian Motion). Application of RW can be found in many other fields: in the percentage increases of stock prices, in

© Springer Nature Switzerland AG 2021
M. Cencini et al., *A Random Walk in Physics*,
https://doi.org/10.1007/978-3-030-72531-0_2

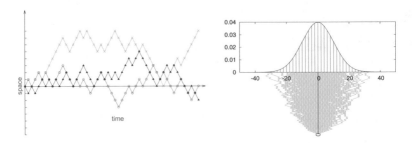

Fig. 2.1 On the left three examples of random path, on the right the limiting Gaussian probability density describing a large number of walker

the exploratory movements of animals in population dynamics, or also in the genetic drift of a population.

In simple cases the walker is limited in his movement in a one-dimensional line or in a two- or three-dimensional space, but it is possible to generalize the path of a walker among different sites (or states). For instance, one can consider a walker performing a journey among N different sites $1, 2, \ldots, N$, which are connected with the following rule: if the walker is in the site i it goes into the site j with a given probability $P_{i \to j}$. If $P_{i \to j} = 0$ there is no direct link between site i and site j. A possible real-world application of such a model describes an aimless journey among villages in a mountainous regions: a non-zero $P_{i \to j}$ means that there exists a road between the village labeled with i and that one labeled with j, and the value of probability $P_{i \to j}$ indicates how dangerous is the road between the two villages, i.e., the closer to zero the probability value the more dangerous the road between the two villages and the walker would not want to face it. Further generalization of RW, in which the walker jumps between connected nodes, can be useful in various and different contexts, from the epidemic spreading to the internet monitoring, not to mention that the idea of RW is at the heart of Google search engine, at least in its first implementation. Moreover, random walk processes are exploited in artificial intelligence algorithms, in which the criteria for decisions are based on testing at random different procedures, and also in the social media context, in which advertisements and suggestions mainly follow the tracked characteristics of the user, but sometimes they need RW to vary suggestions that reach users. Therefore, the random walk, seemingly a mathematical game, is rather a functional instrument in many different theoretical and applied fields.

Finally, discussing the title of the book, we thought the reader can perform a random walk between the various authors and the various aspects of modern physics presented in the book. Clearly some entries are more linked than others, e.g., a connection between Determinism and Laplace sounds more reasonable than Determinism and Volterra. Therefore we suggested few possible paths in the Introduction that can guide the reader to follow and deepen some topic that might be of interest to him. Anyway, we invite the willing reader to look for her/his personal walks.

Chapter 3
Atoms

Surely nowadays nobody doubts about the existence of atoms and that they are the basic building blocks of ordinary matter, this is now a part of the basic scientific knowledge. The enormous relevance of atoms in science has been vividly expressed by R. Feynman in his famous quote: *"If, in some cataclysm, all of scientific knowledge were to be destroyed, and only one sentence passed on to the next generations of creatures, what statement would contain the most information in the fewest words? I believe it is the atomic hypothesis (or the atomic fact, or whatever you wish to call it) that all things are made of atoms-little particles that move around in perpetual motion, attracting each other when they are a little distance apart, but repelling upon being squeezed into one another. In that one sentence, you will see, there is an enormous amount of information about the world, if just a little imagination and thinking are applied"*.

In spite of the great relevance of atoms, until one century ago their real physical existence was rather controversial and even eminent scientists, such as E. Mach, P. Duhem, and W. Ostwald, did not believe that atoms were the basic constituents of the world. Surprisingly even Planck, one of the fathers of modern physics, changed his mind about the physical existence of atoms only at the end of the nineteenth century.

There is no risk to overestimate the relevance of atoms in science. In the following we summarize the development of atomism from its origin up to the modern time. The basic idea of atoms is rather old, as far as we can know it dates back to Leucippus and Democritus (fifth century BC). The Greek adjective "atomos" means "uncuttable"; now we know that atoms have an internal structure (electrons, protons, and neutrons, and, at even finer scales, quarks, strings, etc.). However, this is a marginal aspect that does not alter the relevance of the visionary intuition of the two Greek philosophers about the fact that the world of macroscopic objects is based on the motion of very small entities (atoms) and that the very nature of matter is discrete.

Of course, the atomism in the original formulation of ancient Greeks is rather far from a modern scientific theory, it was mainly a metaphysical thesis whose aim was to establish the ultimate nature of material reality by philosophical arguments.

© Springer Nature Switzerland AG 2021
M. Cencini et al., *A Random Walk in Physics*,
https://doi.org/10.1007/978-3-030-72531-0_3

We cannot resist to cite the famous fragment by Democritus about the distinction between perceived properties like tastes and colors, whose existence is only "by convention", in contrast to the "true" reality, which is atoms and void: *"Sweet exists by convention, bitter by convention, heat by convention, cold by convention, color by convention; but atoms and the void exist in truth"*.

In the old times (say before Galileo and Newton) atomism was not particularly popular, mainly due to the negative opinion about atomism of the two main philosophical schools (of Plato and Aristotle). Moreover atomism, considered a dangerous belief from both Catholic and Protestant Churches, was subject to censorship for a long time. In the fifteenth century a copy of the poem *De Rerum Natura* of the Roman poet Lucretius was discovered by the Italian Poggio Bracciolini in a German monastery. This wonderful poem, which explains the natural world and our place in it in the framework of the atomism of Epicurus (who basically followed Leucippus and Democritus), was considered a dangerous example of atheism and, for several centuries, its reading was forbidden in many universities.

We can say that both Galileo and Newton shared the main ideas emerged in ancient Greece that atoms constitute the ultimate nature of material reality, even if Galileo had not a precise theory of them while Newton thought about atoms in a much more modern perspective.[1] However, the real impact of atomism for science started only in the eighteenth century, and its relevance increased a lot in the second half of the nineteenth century. Roughly speaking, before modern times, the word "atom" has been used with two meanings: what chemists called "chemical atom", that in modern terms corresponds to chemical elements, and what physicists (and more general natural philosophers) indicated as "physical atom", which now corresponds to indivisible particles.

In chemistry, a key step was due to J. Dalton, in the second half of the eighteenth century, with his formulation of chemical atomism: the basic assumption was that chemical elements are composed of "ultimate objects", i.e., atoms; his theory had relevant implications for the way chemicals combine by weight. We can summarize the chemical atomism in the following way: for each element there is a unique indivisible unit which enters into combination with similar units of other elements in small integer multiples. Remarkably, chemical atoms provide a very solid basis to explain the empirical data of stoichiometry, namely, the fact that quantities of reactants and products form a ratio of positive integers, e.g., two moles of nitrogen and one of oxygen (H_2O) are needed to form water.

In physics, the theory of atomism had an empirical (and theoretical) support owing to the kinetic theory of gases which successfully predicted experimental laws, for instance, the celebrated result due to Maxwell that the viscosity of gases is independent of the density. Kinetic theory, which aims to deduce the properties of gases in terms of the collisions between the particles (atoms or molecules) composing

---

[1] *"Have not the small Particles of Bodies certain Powers, Virtues or Forces, by which they act at a distance, not only upon the Rays of Light for reflecting, refracting, and inflecting them, but also upon one another for producing a great part of the Phenomena of Nature?"*, Newton, [Opticks, Book 3, Part 1].

the gas, had its origin with the seminal work of Daniel Bernoulli (first half of the eighteenth century) who was able to explain pressure in terms of the many collisions of atoms with the wall of a vessel. Then the theory had been developed, mainly by Maxwell and Boltzmann (see the related entries), in the second half of the nineteenth century. Kinetic theory was an alternative to the imponderable material called caloric, used until the beginning of the nineteenth century to explain heat. After chemistry and kinetic theory offered explanations of a wide class of phenomena in the framework of atomism, the basic idea was widely accepted and the atomic-molecular hypothesis gained in plausibility for the evidence that gas laws also apply to solutions, which are homogeneous mixtures containing two or more chemical substances, the solute and the solvent.

In spite of these important results, in the second half of the nineteenth century, the success of phenomenological thermodynamics was at the origin of a school (known under the name of "energetics") that rejected atomism. Such an approach, which now may seem surprising, at that time was supported by important scientists and philosophers, e.g., P. Duhem, E. Mach, and W. Ostwald. In their positivistic point of view, the champions of the energetics believed that macroscopic phenomena (such as chemical reactions) should be treated solely in terms of a phenomenological approach based on the conservation of energy and the spontaneous increasing of entropy, and considered atoms just as a useful mathematical tool, with no real ground. In their opinion, since atoms and molecules are invisible, a decisive evidence of the atomic structure was impossible, therefore, the atomic theory must be considered just a physically unverifiable hypothesis and atoms as a mere notion of practical convenience but of no physical reality. The followers of energetics considered phenomenological thermodynamics superior to atomism in the explanation of the second law of thermodynamics. One of the reasons for such an opinion is the fact that Newtonian mechanics, owing to its time reversal property, cannot be exploited to distinguish between past and future, see the entries Boltzmann and Irreversibility.

The controversy about atomism was still open at the end of the nineteenth century: the majority of the British physicists were followers of the atomism, on the contrary energetics was particularly popular in the German scientific community. The main enemy of the energetics was Boltzmann, his confrontations with the opponents of the atomic theory became almost legendary. A rather famous fight took place in September 1895 in Lübeck, at the Congress of German Scientists. Years after, the famous German physicist Sommerfeld described what happened as follows: *"Helm was the champion of energetics; then came Ostwald and, afterward, the philosophical theories of Mach (who was not present at the event). In the opposite corner was Boltzmann, supported by Felix Klein. The skirmish between Boltzmann and Ostwald looked pretty much like a duel between a hefty bull and a trembling bullfighter"*. A remark by A. Einstein on this debate is particularly illuminating: *"the prejudices of these scientists against atomic theory can be undoubtedly attributed to their positivistic philosophical views. This is an interesting example of how philosophical prejudices hinder a correct interpretation of facts even by scientists with bold thinking and subtle intuition"*.

In the early twentieth century, after Einstein and Smoluchovsky theory of Brownian Motion (see related entry) and the experiments of Perrin which validated the theory, even the last exponents of energetics capitulated to the increasingly compelling evidence on the existence of atoms. In 1909, Ostwald did acknowledge that he had been wrong and Arrhenius, another opponent of the atomism, summarizing Einstein's and Perrin's work on Brownian Motion, during a 1911 congress in Paris declared that *"after this we can no longer question the essentiality of the existence of atoms"*. The last one who remained to oppose atomism was Mach, who espoused the lost cause of energetics till the end. Atoms really exist even if we cannot see them (as Mach liked to say), and Perrin had in fact counted them using Einstein's formula on Brownian Motion. As discussed in the entry Brownian motion, the diffusion coefficient, which is accessible experimentally, can be expressed in terms of several known (or measurable) macroscopic quantities (such as the gas constant, the viscosity, the temperature, the radius of the colloidal particle) and, more importantly for atomism, the Avogadro number N_A.[2] The validity of the relation between the diffusion coefficient and macroscopic quantities allowed for a decisive conclusion about the existence of atoms; in the words of Einstein *"if the prediction of this motion were to be proven wrong, this fact would provide a weighty argument against the molecular-kinetic conception of heat"*.

The value of the Avogadro number N_A measured by Perrin was later confirmed by different measurements not directly tied to kinetic theory. For instance, Rayleigh speculated that the color blue of the sky was due to the scattering of sunlight with gas molecules present in the atmosphere, rather than suspended particles (such as water drops).[3] In a way rather similar to the case of the Brownian Motion, Rayleigh scattering theory allows to establish a relation between the Avogadro number and quantities which can be measured in optical experiments. Remarkably the value of the Avogadro number obtained with such an approach is in good agreement with the one found in Perrin experiments on Brownian particle diffusion.

In the twentieth century, with the progress of quantum physics, nuclear physics, modern chemistry, and so on, we had a complete triumph of the atomism. Beyond the results on physical reality of atoms from Brownian Motion and light scattering, we can mention many other clear evidences in the modern physics, e.g., phenomena as ionization, cathode rays, and radioactive decay. In addition, very important from a theoretical point of view, the periodicity of the relations between the properties of the elements and their atomic weight indicates in a clear way that the atoms are composed of smaller particles (electrons and nuclei). The accurate description of the behavior of elements at varying their atomic weight is surely one of the great success

---

[2]The Avogadro number is the number of elementary entities (atoms or molecules) comprising one mole of a given substance, for instance, the number of atoms present in 12 g of isotopically pure carbon-12, its numerical value is 6.0221415×10^{23}.

[3]The mechanism proposed by Rayleigh, now called Rayleigh scattering, which is responsible for the blue of the sky, involves the scattering of light with particles smaller than the wavelength of the light, while scattering with particles larger than the wavelength of light (called Mie scattering) is responsible, e.g., of the red color of the sunset, which can be very intense over the sea where a lot of water vapor droplets and small aerosol particles are typically present.

of quantum mechanics as well of atomism. A discussion of some of these issues and, most importantly, of the many discoveries occurred when studying the internal structure of atoms, can be found in the entry Fermi.

We conclude by mentioning that, nowadays, we can actually "see" the atoms and the ultimate discrete structure of the matter, for instance, with the help of scanning tunneling microscopy we can observe real-space images of surfaces at atomic-scale resolution.

Chapter 4
Big Data

Usually it is assumed that the research activity of the hard sciences (such as physics and chemistry) is based on three pillars, namely, theory and experiments (which are the traditional ones, say since Galilei) to which during the twentieth century it has been added the use of numerical computations (see entry Computer, algorithms and simulations), which nowadays is recognized as the third pillar. In the last decades, at least according to some authors, we had the rise of a fourth paradigm which is data mining, i.e., the exploration of a large amount of data through powerful tools of analysis.

As a precursor of such a trend we can mention that in the 1980s some researchers in the field of artificial intelligence (AI) devised BACON (after the British philosopher Francis Bacon), a computer program "able" to automate scientific discoveries. Apparently the program "discovered" some physical laws, including Kepler's third law. It is interesting to look at the details of the procedure used by BACON. The program received as input the numerical values of the distances from the Sun, D, and the revolution periods, P, of planets. BACON, then, discovered that D^3 is proportional to P^2. While this is surely interesting, we tend to think that it is difficult to claim that this represents a direct inductive approach only from data. Actually for Kepler the raw observables were not D and P, but a huge list of planetary positions seen from the Earth at different times, i.e., something much more difficult to interpret than the (clean) data given to BACON. Remarkably in his discovery, Kepler, who was guided by strong beliefs in mathematical harmonies as well as the controversial (at that time) heliocentric theory of Copernicus, was able to guess the "right" variables D and P.

More recently, some scientists trained an algorithm that, using a learning strategy just based on the knowledge of the past evolution of the system, succeeded in discovering the laws of motion of a chaotic double pendulum, which is a rather complex system. Therefore, despite being given no scientific knowledge by the researchers, the machine had, apparently, deduced certain not trivial laws of physics. Actually examples of this kind are nowadays more and more reported in the scientific literature,

© Springer Nature Switzerland AG 2021
M. Cencini et al., *A Random Walk in Physics*,
https://doi.org/10.1007/978-3-030-72531-0_4

fueling an increasing widely shared belief that the explosive rise in technology for data gathering and analysis may soon make the scientific method unnecessary. The message is clear: forget models and hypotheses, all we will need are smart machine-learning algorithms devouring huge datasets. Through fully automated learning, we will come to make vastly more accurate predictions, and face fewer real-world surprises. Machines will do science for us.

In 2008 Chris Anderson (the chief editor of the influential magazine *Wired*) wrote a paper with a rather explicit title *The End of Theory: The Data Deluge Makes the Scientific Method Obsolete*. Such a paper quickly became a sort of ideological manifesto of the enthusiasm for (model-free) data science. The basic idea, using Anderson's words, is that *"Correlation is enough"*: now, in the age of Big Data, with the possibility to have an enormous amount of data, we can stop looking for models and start analyzing data without an understanding of the basic mechanisms ruling the phenomenon.

However, while the concept of correlation is crucial in many physical theories, correlations without an underlying solid theory can be "spurious" and thus misleading. Indeed, in general, the presence of correlations does not imply a causal dependence between different phenomena. For instance, in the book *Spurious Correlations* by Tyler Vigen (2015), it is possible to find a lot of hilarious examples, we mention just a couple of them: between the "US spending on science, space and technology" and "Number of suicides by hanging, strangulation and suffocation" (correlation coefficient 99.79%), or between the "trend of divorces in Maine" and "the per capita consumption of margarine in U.S.A" (correlation coefficient 99.26%).

The enthusiasm toward artificial intelligence approaches and science based only on data actually goes much further. For example, some computer scientists believe one day we may find "the master algorithm", an ultimate data-processing device that, running quite independently from human oversight, will organize human politics to "speed poverty's decline" and to allow to live "longer, happier and more productive" (see the best seller of P. Domingos, *The Master Algorithm*, 2015).

However, it is worth expressing a few cautionary comments concerning a naive approach only based on Data even if Big. One of the most important aspects of science is surely forecasting, and in a quite natural way political institutions make substantial use of forecasting to devise their policies, e.g., in economy. Interestingly in such a topic the boundaries between natural and social sciences are often crossed, as well as the boundaries between the scientific, technological and the ethical domains. For instance, most of us rely on weather forecasts to plan the daily activities. Once one accepts a Big Data point of view, we can be dispensed with theory, or even with hypotheses, we will just concede the assumption of having a large amount of data, combined with powerful algorithms, to forecast a deterministic (see the entry Determinism) or, at least, quite regular system. Then, the basic reasoning would be that "Similar premises lead to similar conclusions", in Maxwell's words *"That from like antecedents follow like consequents"*.

In order to investigate the severe limitations of a purely inductive[1] approach in science, even in a deterministic context, we now briefly discuss a natural Big Data approach to the forecasting problem, that is using just the knowledge of the past without the aid of theory. Basically, one looks for a past state of the system which is "close" to the present one: if it can be found at day k of the past, then it makes sense to assume that tomorrow the system will be "close" to day $k + 1$ of the past. In more formal terms, given the series $\{\mathbf{x}_1, \ldots, \mathbf{x}_M\}$ where $\mathbf{x}_j$ is the vector describing the state of the system at time $j \Delta t$, we look in the past for an analogous state, that is a vector $\mathbf{x}_k$ with $k < M$ that is "close enough" (i.e., such that $|\mathbf{x}_k - \mathbf{x}_M| < \epsilon$) being ϵ the desired degree of accuracy). If we are able to find such a vector, then we can "predict" the future at times $M + n > M$ by simply assuming for $\mathbf{x}_{M+n}$ the state $\mathbf{x}_{k+n}$.[2] It all seems quite easy, but the subtle point is in the following questions: which is the proper vector $\mathbf{x}$ which describes the state of the system we want to predict (i.e., for instance, to characterize the state of an ideal gas we need to know temperature, density and pressure) and more crucially how difficult is to find a close enough analog to the present state, which we can use to make the forecast?

Finding the proper variables, as the history of physics teaches us, is quite non-trivial, but even assuming that we know them the problem of finding an analog is quite subtle. The possibility to find analogs is strictly linked to the celebrated Poincaré recurrence theorem: after a suitable time, a deterministic system with a bounded phase space returns to a state arbitrarily close to its initial one. This sounds good, as it implies that analogs surely exist. Now the delicate question is: how long have we to go back to find it (that is how much data should we collect)? The answer has been given by the Polish mathematician Mark Kac[3]: the average return time in a region A is proportional to the inverse of the probability, $P(A)$, that the system is in A. To understand the difficulty to observe a recurrence, and hence to find an analog, consider a system of dimension D,[4] the probability $P(A)$ of being in a region A that extends in every direction by a fraction ϵ is proportional to ϵ^D, therefore the mean recurrence time is $t_r \sim \epsilon^{-D}$. If D is large (say, larger than 10), even for not very high levels of precision (for instance, 5%, that is $\epsilon = 0.05$), the return time is so large that in practice a recurrence is never observed (or, equivalently, an analog cannot be

---

[1] About the (ab)use of inductivism we cannot resist quoting Bertrand Russell *"Domestic animals expect food when they see the person who usually feeds them. We know that all these rather crude expectations of uniformity are liable to be misleading. The man who has fed the chicken every day throughout its life at last wrings its neck instead, showing that more refined views as to the uniformity of nature would have been useful to the chicken".* from *The problems of Philosophy* (1912).

[2] If we are lucky enough to find several "close" analogs, we can, for instance, suitably average the predictions obtained by each of them to improve the forecast.

[3] Actually a version of this result was already derived by Boltzmann who used it in his hot debate about irreversibility with Zermelo.

[4] To be precise, D is the number of variables (often called degrees of freedom) needed to unambiguously determine the state of the systems, and if the system is dissipative, D is the fractal dimension of the attractor, see the entry Chaos or Fractals.

found even with a huge number of historical data): this is often referred to as the *curse of dimensionality*.

The above difficulty due to the dimensionality is only a part of the problem. Usually we do not know the vector **x** describing the state of the system. Such a difficulty is rather serious and is well known in statistical physics; it has been stressed, e.g., by Onsager and Machlup in their seminal work on fluctuations and irreversible processes, with the caveat: *"how do you know you have taken enough variables, for it to be Markovian?"*[5]; similarly Ma says *"the hidden worry of thermodynamics is: we do not know how many coordinates or forces are necessary to completely specify an equilibrium state"*.

Since often there is not a sure protocol to individuate the proper variables, to describe the state of the system, one is forced to use just the time series $\{u_1, \ldots, u_M\}$, where u_j is an observable sampled at the discrete times $j\Delta t$. Now the obvious question is: how to find the proper variables from the time series? The mathematician Takens had given an important contribution to answering the above question. He showed that from the study of a time series $\{u_1, \ldots, u_M\}$, it is possible (under some technical considerations, e.g., if we know that the system is deterministic and it is described by a finite-dimensional vector, and M is large enough) to reconstruct the system dynamics in terms of a state vector equivalent to the original variables **x**. Unfortunately, at practical level, the method has rather severe limitations:

(a) It works only if we know a priori that the system is deterministic;
(b) The protocol fails if the dimension, D, of the attractor (see entry Chaos) is large enough (say more than 5 or 6).

Once again the main difficulty is a consequence of Kac's lemma: the minimum size of the time window M allowing for the use of Taken's theory increases as C^D where C depends on the wished accuracy, we can assume that for a fair result, say 1%, we have to take C in the order of 100 per degree of freedom. Therefore, the method cannot be used to build up a model from data, apart for special cases (when the system is low dimensional, i.e. its states are specified by a small number of variables). One of the few successes of the method of the analogs is the tidal prediction from past history, in spite of the fact the tides are chaotic. The reason for analogs to work in the case of tides is the low number of effective degrees of freedom involved. Therefore, the most serious limit to predictions based on analogs is not the sensitivity to initial conditions, typical of chaos, but, as realized by Lorenz in 1960s, it must be charged to the high dimensionality of the systems.

An analysis of weather forecasts provides a very good illustration of the main aspects of predictive models. Not least because of the extreme accuracy which this field managed to achieve over the past decades that has been attained only when it became clear that *too much data* would be detrimental to the accuracy of the model. Indeed, as we now briefly review, in the early days weather forecasts featured a naive form of inductivism not dissimilar to the one fueling the Big Data enthusiasm.

---

[5]Being Markovian, without entering into many details, implies a great simplification in the description of the phenomena, as it means that the future state of a system just depends on the present one, even if not deterministically.

The first modern steps in weather forecasting are due to the British scientist Lewis Fry Richardson who, in his visionary book *Weather prediction by numerical process* (1922), introduced many of the ideas on which modern meteorology is based. His approach was, to a certain extent, in line with a genuine reductionism (see entry Determinism), trying to predict weather evolution by numerically solving the partial differential equations for the hydrodynamic and thermodynamics of the atmosphere, with initial conditions given by the present state of the system. However, while the key idea of Richardson was correct, realizing it in practice requires an enormous computational power, not available even today. Therefore it is necessary to deeply simplify the model (exploiting a number of non-trivial assumptions and approximations) by introducing effective equations focusing only on the essential elements of the dynamics of the atmosphere and ignoring all those high frequency components that, irrelevant for meteorology, deeply impact on the numerical feasibility and stability. Hence a less accurate model built on solid physical insights works incredibly better for weather forecast than a more precise model. For an accurate discussion see entries Richardson and Laws, levels of description, and models.

The above cautionary words should not be interpreted as a suggestion to give up with believing in, or collecting, Big Data. Indeed it is certainly undeniable that Big Data associated with artificial intelligence algorithms have revolutionized entire technical sectors, such as automatic image recognition, automatic translation, speech recognition, self-driving cars, and so on. All classification problems have received incredible benefits from the use of machine learning applied to Big Data and it is also true that machine learning is impacting and, possibly, will impact even more science. However, in our opinion, the role of Big Data in science will not make modeling outdated and theoreticians jobless. In other terms we should avoid believing or pretending that Big Data is magic, but it must be considered for what it is, i.e., a new very powerful investigation tool to be used in appropriate contexts and which can surely empower scientific research.

In our opinion science must revolve, fundamentally, around the building of models that describe our world. The elements and the structure of models give a deep understanding of phenomena contrary to the knowledge provided by algorithms of artificial intelligence applied to Big Data which often act as black boxes without strong descriptive and interpretative capacity. For instance, in order to understand the relationship between smoking and cancer, we need to run experiments and develop physiological and biochemical understandings of the interaction between oncogenes, epigenetics, and inflammation. Merely tabulating a gigantic database of every smoker and nonsmoker in every city in the world, with great detail about when they smoked, where they smoked, how they lived, and how they died would not be enough, no matter the dimension of the collected data file, to deduce all the complex underlying biological machinery. In this respect we do not consider outdated the thoughts of Poincaré: *"The Scientist must set in order. Science is built up with facts, as a house is with stones. But a collection of facts is no more a science than a heap of stones is a house"*.

Chapter 5
Boltzmann

Ludwig E. Boltzmann (1844–1906) was born in Vienna on February 20, the night between Carnival Tuesday and Ash Wednesday; he used to say, jokingly, that that was the origin of his changes of mood, which swung from moments of sheer enthusiasm to periods of severe depression. He committed suicide in Duino (near Trieste) on September 5, 1906. Ludwig Boltzmann (LB) was a restless soul, a generous scientist looking for a tranquility of the mind that he could never find. He moved from one university to the next relentlessly: in 1869 he was appointed professor in Graz, in 1873 he transferred to Vienna, then in 1876 he went back to Graz. Few years later he accepted a prestigious chair in Berlin, but he changed his mind and never took it. In 1890 he went to Munich, in 1894 to Vienna, in 1900 to Leipzig. At the end he went back to Vienna once again in 1902, at this point the Emperor Franz Josef, annoyed by this erratic behavior, required a formal written statement that he would not move yet another time.

Despite the many awards and recognitions he received, Boltzmann often felt isolated and unappreciated, at least toward the end of his life. In the preface to his most important book *Lectures on Gas Theory*, he wrote *"I am conscious of being only an individual struggling weakly against the stream of time. But it still remains in my power to contribute in such a way that, when the theory of gases is again revived, not too much will have to be rediscovered"*.

In the scientific debates LB was a strong fighter, and his confrontations with the opponents of the atomic theory became legendary. The most famous dispute took place in September 1895 in Lubeck, at the Congress of German Scientists. From one side the champions of energetism (in particular Ostwald and Mach) to the opposite corner was Boltzmann, supported by Felix Klein. The vivid contention between Ostwald and Boltzmann, who nevertheless remained friends, carried on in a series of papers. Only in 1909 did Ostwald eventually acknowledge that he had been wrong, while Mach never backtracked.

Among the non-technical works of LB we like to mention the nice booklet *Journey of a German professor in the Eldorado* an amusing description of his trip to California

© Springer Nature Switzerland AG 2021
M. Cencini et al., *A Random Walk in Physics*,
https://doi.org/10.1007/978-3-030-72531-0_5

(1905). A significant part of the book is devoted to food and beverages, as apparently the Californian cuisine was not to his liking. Boltzmann admired the American way of life, which was freer and much more democratic than the lifestyle in the Habsburg empire, but he could not stand the puritanism: *"They hide the wine, as students do with their cigars. This is what they call freedom"*. He was disappointed to see that anyone asking for a wine store was severely frowned upon, as if he had enquired about *"those ladies"*.

LB's scientific career was mainly devoted to the building of a statistical theory to describe the thermodynamics behavior of macroscopic systems starting from their microscopic components and laws, but there are also other lesser known yet far from negligible results. At the beginning of his career LB worked a lot on electromagnetism. He was among the first who understood the importance of Maxwell's equations, and the connection between optics and electromagnetism, in particular the relationship between the electric constant and the refraction index. Following the famous experiments of Hertz on electromagnetic waves, LB became so enthusiastic about the subject that in 1886 he repeated himself the experiments, gave courses on electromagnetism and published his lectures (in two volumes).

In 1872, LB published what is universally known as the Boltzmann equation, which governs the evolution of the probability density of the velocity in rarefied gases. The Boltzmann equation represents one of the most elegant and profound chapters of theoretical physics; it allowed to recover the Maxwell (Boltzmann) distribution on a purely dynamical ground and to formulate the celebrated H-theorem. Remarkably, it marked the first appearance in the scientific literature of an evolution equation for a probability density and, possibly, the first example in which probability plays a truly foundational role in the natural sciences. Besides its theoretical importance, the Boltzmann equation has seen a host of applications in many scientific and technological fields, from semiconductors to the dynamics of rarefied gases in aerospace engineering. His contribution to statistical mechanics culminated in 1877, when LB published the probabilistic interpretation of thermodynamics, incorporated in the celebrated formula $S = k_B \log W$ engraved on Boltzmann's tombstone (see Fig. 5.1).

Boltzmann was also one of the fathers of hereditary mechanics, that is, the study of the effects of memory of past events on the deformations of certain materials, such as glasses (1874). These contributions show up in integral equations that will be developed by Picard and Volterra only at the beginning of the twentieth century.

In 1884, he proved that the total energy of a black body is proportional to the fourth power of the temperature (the so-called Stefan-Boltzmann's law), thus confirming the intuition of his teacher Stefan, based on a few experimental observations. Although from a technical point of view the derivation of the Stefan-Boltzmann's law is quite simple, at that time, it was conceptually rather important in showing that thermodynamics does not hold just for gases.

In the final decade of the nineteenth century, trying to answer the criticisms raised by Zermelo and Loschmidt, Boltzmann perfected his theory of statistical mechanics and irreversibility and wrote the monumental opus *Lectures on Gas Theory*.

$$S = k \log W$$

Fig. 5.1 Boltzmann's tombstone in Vienna. Photograph: the authors of this book in 2007

He had a great admiration for Darwin, and considered the theory of evolution as the most important discovery of the nineteenth century. Boltzmann was very interested in philosophy as well, albeit not in a systematic way (his main fascination clearly remained with physics) and in the final years of his life he gave a course on philosophy of science. Some of his ideas are extremely intriguing and anticipating those of Thomas Kuhn on scientific revolutions and paradigms: *"The man on the street might think that new notions and explanations of phenomena are gradually added to the bulk of the existing knowledge (...). But this is untrue, and theoretical physics has always developed by sudden jumps (...). If we look closely into the evolution process of a theory, the first thing we see is that it does not go smoothly at all, as we would expect; rather, it is full of discontinuities, and at least apparently it does not follow the logically simplest path".*

Sometimes LB has been misunderstood. For example, Karl Popper, albeit admiring Boltzmann, accused him of having constructed theories that could not be falsified and called him an idealist. The criticism refers specifically to Boltzmann's computation of the return time of a macroscopic system. For macroscopic objects, where the number of particles is gigantic, in the order of Avogadro's number 6×10^{23}, the computation gives a time much longer than the universe's age, which according to Popper was a non-falsifiable hypothesis: *"I think that [Boltzmann's idea of irreversibility] is completely unsustainable, at least by a realist. It presents the unidirectional evolution as an illusion (...). So it transforms our world into an illusion, together with all our efforts to understand the world better. But eventually it defeats itself (as any form of idealism)"*. The accusation is unfair, and also technically wrong: Boltzmann's intuition on return times was rigorously proved by Mark Kac in 1947. Moreover, if we consider small systems, the problem can be numerically analyzed using a computer and Boltzmann's prediction is in perfect agreement with the actual results.

The dominant idea in Boltzmann was the attempt to find a bridge between the (microscopic) laws of mechanics and the (macroscopic) laws of thermodynamics. A challenging problem both conceptually and practically, as it requires linking two worlds with very different features even at a qualitative level, to mention just the most eye-catching: mechanical laws are reversible while thermodynamics is blatantly irreversible as stated by the second law of thermodynamics. In the mechanical description of a macroscopic object one has to face with Newton's equations for the evolution of the positions and velocities of a huge number of particles (the body constituents). Conversely, in thermodynamics one deals just with few global qualities, e.g., temperature and pressure, characterizing the macroscopic body.

The two milestones of long and winding road which eventually led to Boltzmann's grand vision are: the introduction of probabilistic ideas and their interpretation in physical terms; the relationship which links the macroscopic world (thermodynamics) to the microscopic one (dynamics).

The former one is extremely delicate and it is the object of intense study still today. LB's idea was to substitute time averages with averages over a suitable probability density. This conjecture is called the ergodic hypothesis. If proved true, the probability of a region A in phase space[1] is nothing but the fraction of time spent in A during the evolution, computed over a very long time. The ergodic hypothesis allows for the introduction, in a consistent way, of probabilistic concepts (the so-called statistical ensembles), in deterministic systems and, which is the key point, allows for the possibility of explicit computations.

The latter milestone is engraved on his tombstone, $S = k_B \log W$, and connects the entropy, S, of the macroscopic body (a thermodynamic quantity) to the (logarithm of the) number of microscopic states, W, (a mechanical-like quantity) with which a given macroscopic configuration can be obtained; k_B is Boltzmann's constant. This law is one of the greatest achievements of science (those like $F = ma$ and $E = mc^2$, which rightly become pop enough to be found on some T-shirts) and subsumes a large part of statistical mechanics.

Surely the most important (and controversial) contribution of LB to science has been his theory of irreversibility (H-theorem). On the one hand, in everyday life we experience phenomena clearly indicating that natural processes involving macro-

---

[1] That is the huge dimensional space made of the 3N positions and 3N velocities of the N microscopic constituents of a gas, i.e., molecules or atoms.

scopic systems are intrinsically irreversible. On the other hand, Newton's equations, ruling the microscopic components of macroscopic systems, are invariant (i.e., symmetric) with respect to the time reversal transformation: if at a certain time t we change the sign of the velocities of the particles the system traces back its history and a time t after the "reversal" it returns to the same initial position, with the velocities reversed. Such a property is at the origin of the very difficult (apparently without solution) problem of the irreversibility: how can we reconcile the macroscopic irreversibility with the reversible nature of the dynamics? In the entry irreversibility we briefly discuss Boltzmann's answer to the above question.

The greatness of Boltzmann's work comes not so much from having "reduced thermodynamics to mechanics", as some authors claim; rather, it lies in having realized that irreversibility cannot be understood solely with the laws of mechanics. He made us comprehend the subtle nature of the emergence of irreversibility, for which two fundamental ingredients are necessary:

• a large number of particles (atoms or molecules), which is at the heart of the substantial imbalance between the microscopic and macroscopic scales;

• suitable initial conditions (those generating molecular chaos): the laws of mechanics have nothing of the kind of the second law of thermodynamics, which can be reduced to mechanical terms only by means of special assumptions on the initial conditions.

We may add to these two items a third element, somehow related, and complementary to the second one, namely, the use of probability: not all microscopic states evolve in an irreversible way, but only "most of them". With macroscopic systems, where the number of particles is enormous, it is practically certain that an irreversible behavior will be seen.

Although Boltzmann's stature is undoubted, yet about 180 years after his birth, there still are historically inaccurate accounts of his life and contributions to science. For instance, the introduction of statistical ensembles, often attributed to Gibbs, was actually conceived by Boltzmann. There is no shortage of erroneous opinions, some completely misleading, e.g., Prigogine and Stenger wrote on many occasions that Boltzmann's ideas were incoherent, if not utterly wrong. Nothing is farther from the truth: Boltzmann's intuitions have been systematically confirmed by a number of rigorous mathematical papers and numerical simulations. We can safely state that today even the most controversial aspects of the approach of LB to irreversibility have been clarified, thus showing that his intuitions were eventually correct. In particular, thanks to the rigorous mathematical results on many scientists (e.g., Grad, Illner, Lanford, Shinbrot, Di Perna, Lions, Pulvirenti, Cercignani) we know that the H-theorem is correct, at least in a suitable limit.[2]

During his life Boltzmann did not manage to convince all his critics that atoms exist as physical entities, and not just as a useful hypothesis for computations, as Mach believed and wrote *"the atomic theory plays in physics a role similar to certain mathematical concepts; it is a model to help figuring out certain facts"*. Now we

---

[2]Specifically, theorems have been proved when the number of particles per unit volume, N say, tend to infinity while their size σ tends to 0, in such a way that $N\sigma^2$ is constant.

know that atoms really exist even if we cannot see them (as Mach liked to say), and Perrin, with his experimental work, had in fact counted them using Einstein's formula on Brownian Motion.[3] Apparently LB was not aware of Einstein's 1905 work on Brownian Motion, which can be considered as the triumph of Boltzmann's idea about atoms and statistical mechanics. It is interesting that Einstein did not care about the Brownian Motion in itself, his real purpose being to show that bodies visible under the microscope and suspended in a liquid possess a movement in agreement with the molecular theory of heat, making the effects of molecules observable.

LB's scientific legacy is gigantic: his ideas were the starting point for whole new areas of mathematics, physics, and technology, e.g., the ergodic theory, the theory of large fluctuations, and transport phenomena. The work of LB influenced Planck's study of black bodies that eventually led to the development of quantum mechanics, and provided the foundations of the modern techniques of numerical simulations. For instance, Boltzmann's equation, originally introduced for diluted gases (see the entry Irreversibility) is today widely used in numerical simulations of technological relevance, such as aerospace engineering.

We like to conclude briefly discussing the contribution of LB to ergodic theory. The ergodic hypothesis, in the original formulation proposed by LB, is, strictly speaking, wrong.[4] However, LB had the merit to give the birth to ergodic theory which played and still plays a crucial role in the development of statistical mechanics. First of all, it allows for introducing probability in the framework of deterministic systems in a natural way. In particular, as ergodicity is equivalent to the assumption that different trajectories have the same asymptotic features, it provides the means to interpret probability in frequentist terms. Assuming ergodicity, the probability of an event (defined via its asymptotic frequency) is an objective and measurable property: it suffices to follow a trajectory long enough. It is quite natural to assume that a proper statistical approach can be obtained by time averages taken along the time evolution of the system. This is what actually it is done using numerical simulations, e.g., with the molecular dynamics method, widely used to find the thermodynamics properties of macroscopic objects: basically, invoking the ergodic hypothesis, it is possible to determine the relevant quantities for thermodynamics in terms of temporal averages computed following the evolution of the system by numerically solving the Newton's equations.

---

[3] A macroscopically tiny but microscopically huge object like a pollen grain, when immersed in a liquid, zig-zags in a highly irregular way: the motion has very precise properties statistically speaking (see the entry Brownian Motion).

[4] Indeed, as proved by the KAM theorem, as well the seminal numerical computations by Fermi (see the entries devoted to Kolmogorov and Fermi), strictly speaking, a typical mechanical system is not ergodic, see also the entry Statistical Mechanics.

Chapter 6
Brownian Motion

In 1827, the Scottish botanist Robert Brown while studying an Australian plant (*Clarkia Pulchella*) discovered a phenomenon, now called Brownian Motion (BM), which would have played an important role in physics. Brown was an important scientist, he is credited with the first clear description of the cell nucleus; moreover, he had a formidable reputation as an expert of microscopes. Many scientists sought his advice on microscopes; the most famous was the young Darwin, when he was preparing his voyage on the Beagle. Nowadays Brown is much better known in physics (as well as in mathematics and finance) for the discovery of the BM.

He observed under a microscope that a pollen grain—a few micron object which is small at macroscopic level, but large compared with water molecules—suspended in water displays a rapid and irregular movement, which seemed to continue forever. In Fig. 6.1 it is shown a typical trajectory of a Brownian Motion obtained via a computer simulation. At first, he thought that the phenomenon was linked to the organic nature of pollen grains. However, he soon ruled out such a hypothesis by observing that the same phenomenon occurred even for small grains of inorganic matter. At the beginning the physicists considered such a phenomenon as a sort of curiosity. After some decades, however, the relevance of Brownian Motion and its connection with thermodynamics was realized. For instance, the Italian physicist Cantoni in 1868 claimed that Brownian Motion is a *"beautiful and direct experimental demonstration of the fundamental principles of the mechanical theory of heat"*. In 1888, Gouy pointed out that BM seems to violate the second law of thermodynamics (see the entry Entropy), because the thermal (molecular) agitation of water was converted into mechanical work by moving the suspended particle. He also was among the first to realize that BM was a "natural laboratory" to understand how kinetic theory and thermodynamics could be reconciled. On the other hand it is remarkable that the founders of statistical mechanics and kinetic theory (Clausius, Maxwell, Boltzmann, and Gibbs) did not show interest for the BM (Fig. 6.1).

© Springer Nature Switzerland AG 2021
M. Cencini et al., *A Random Walk in Physics*,
https://doi.org/10.1007/978-3-030-72531-0_6

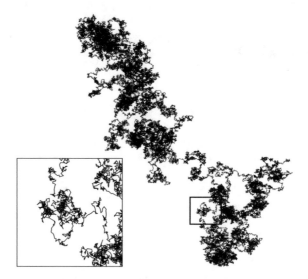

Fig. 6.1 Typical trajectory of a Brownian particle, the zoom shows the self-similar characteristic of these kind of motions

Some scholars (e.g., Truesdell) claim that the first description of the BM is due to Lucretius who, in his poem *On the Nature of Things*, discussed in a remarkable way the motion of dust particles, and used this to support the ideas of Leucippus and Democritos on the existence of atoms:

> Observe what happens when sunbeams are admitted into a building and shed light on its shadowy places. You will see a multitude of tiny particles mingling in a multitude of ways... their dancing is an actual indication of underlying movements of matter that are hidden from our sight... It originates with the atoms which move of themselves. [...] So the movement mounts up from the atoms and gradually emerges to the level of our senses, so that those bodies are in motion that we see in sunbeams, moved by blows that remain invisible.(verses 113-140 Book II)

Likely the motion described by Lucretius is a macroscopic phenomenon due to the movement of air. Nevertheless his main idea about atomism was correct and his words perfectly describes the physics of BM.

It is fair to say that at the beginning of the twentieth century several scientists (we just quoted Cantoni and Gouy, but there were several others) realized that Brownian Motion should have some deep link with kinetic theory and atomism, but none had yet found the proper quantity to look at. The man who fully realized the relevance of the Brownian Motion and understood how to exploit it—i.e., what to measure— to eventually demonstrate the atomic structure of matter was Albert Einstein (see related entry). In his *annus mirabilis* (1905), Einstein put forward his revolutionary theory, and published a few other works on the subject up to 1908. In the same years, independently, also Smoluchowski was thinking of BM but published his work a year later, in 1906. In 1908, Langevin elaborated further Einstein theory in terms of stochastic differential equations (using modern terminology).

Einstein was aware of the kinetic theory of gases developed by Maxwell and Boltzmann. While accepting that individual atoms could not be visible, owing to their small size and huge speeds, he was the first to realize that if kinetic theory and statistical mechanics were correct then any small (but finite-size, and thus visible) particle immersed in a fluid (i.e., a thermal bath of atoms or, more properly for water, of molecules) should behave as a sort of "big" molecule (or atom). In other terms, thanks to the collisions with the molecules, it should move and acquire a kinetic energy equal to $k_B T/2$ (where k_B and T are the Boltzmann constant and the fluid temperature, respectively) for each degree of freedom characterizing its motion. This is, indeed, the essence of the *equipartition of energy* which stands at the basis of statistical mechanics (see the related entry). Therefore, the never stopping motion of the pollen grain is just the fingerprint of the thermal agitation of fluid molecules and no violation of the second law of thermodynamics should be invoked.

In 1906, the above idea was further clarified by Smoluchowski, who directly used kinetic theory to perform the calculations. In this respect, the importance of Smoluchowski was to have spelled out that, although the single collision of the grain with a molecule would be not enough to generate movement, owing to the large number of fast and small molecules, the number of collisions is extremely large and can generate motion. A counter argument could be that the collisions are expected from all directions and so there should be no net movement. However, as realized by Smoluchowski, there will always be unbalance due to fluctuations in the order of the square root of the number of collisions, still a number large enough to cause (detectable) movement.

The other ingredient of Einstein theory was a classical result from hydrodynamics, namely, the Stokes law for the frictional force exerted on a body moving in a liquid: a spherical body of radius a moving with velocity v in a fluid of viscosity η is slowed down by a force given by $-6\pi \eta a\, v$.

Once the phenomenon is formalized in this way, he derived that the particle performs an irregular motion, indeed very similar to the random walk described in the first entry of this book, characterized by the fact that the square displacement of the particle, $\Delta x(t) = |x(t) - x(0)|^2$ (where $x(t)$ denotes the position of the particle at time t), on average grows linearly in time:

$$\langle |\Delta x(t)|^2 \rangle \simeq 6Dt,$$

which is the Einstein-Smoluchowski law of diffusion. Then, using energy equipartition and the Stokes law, he derived that the diffusion coefficient can be expressed as

$$D = \frac{k_B T}{6\pi \eta a} = \frac{RT}{6N_A \pi \eta a}.$$

In the above expression, the second equality stems from the fact that the Boltzmann constant can be expressed as the ratio of the constant of the gas R and the Avogadro number N_A, which at that time was credited to be the number of atoms or molecules contained in a mole (i.e., the mass in grams of a chemical compound/element that equal the molecular mass expressed in Daltons of the compound/element),[1] even though atomism was still not accepted.

The above formula of the diffusion coefficient represents the much desired link between the microworlds of atoms—the Avogadro number N_A—and the macroworlds (or better mesoscopic, see the entry Mesoscale Systems)—the suspended particle. The Einstein-Smoluchowski diffusion law, however, is key to exploit the formula as it provides the much desired quantity to measure, namely, the average square displacement of a suspended particle, from which the diffusion coefficient can be estimated and, in turn, the Avogadro number derived, as the other quantities were either known or easy to measure.

In his biography, Einstein clearly exposed his motivation for studying the Brownian Motion: *"My major aim in this was to find facts which would guarantee the existence of atoms"*. To prove (or disprove) the existence of atoms was indeed Einstein's purpose, as clearly stated in the second paragraph of his paper: *"If the movement discussed here can actually be observed (together with the laws relating to it that one would expect to find), then classical thermodynamics can no longer be looked upon as applicable with precision to bodies even of dimensions distinguishable in a microscope: an exact determination of actual atomic dimensions is then possible. On the other hand, had the prediction of this movement proven to be incorrect, a weighty argument would be provided against the molecular-kinetic conception of heat"*. Moreover Einstein mentioned his own astonishment at the fact that this result had not been obtained by Boltzmann: *"it is puzzling that Boltzmann did not himself draw this most perspicuous consequence, since Boltzmann had laid the foundations for the whole subject"*.

After the theoretical work of Einstein (and Smoluchowski), Svedberg performed some experiments on the diffusion of colloidal particles, but his results were generally confused and wrongly interpreted. The conclusive experimental contribution was by Jean Baptiste Perrin (1870–1942) with his study on the sedimentation and the diffusion of tiny particles in water. The agreement between theory and experiments earned Perrin the Nobel Prize in 1926 *"for his work on the discontinuous structure of matter, and especially for his discovery of sedimentation equilibrium"*. In a lecture in Paris, in 1911, Arrhenius, summarizing the works of Einstein and Perrin, declared *"after this, it does not seem possible to doubt that the molecular theory entertained by the philosophers of antiquity, Leucippus and Democritos, has attained the truth at least in essentials"*. Since then atoms could be "counted" and "measured".

The determination of the Avogadro number, N_A, from the measurement of the diffusion coefficient, D, and its agreement with values obtained from independent measurements definitively closed the heated controversy about the existence of atoms between Boltzmann and the main exponents of energetics, Mach and Ostwald, who considered the atoms useful but non-real for the building of a consistent description

---

[1] We recall from elementary thermodynamics of ideal gases that, by denoting with P, V, T, n the pressure, volume, temperature, and the number of moles of a gas, it holds the equation of state $PV = nRT$, which can be rewritten as $PV = nN_A k_B T$, with $N = nN_A$ denoting the total number of molecules/atoms in the gas.

of nature: in Mach's words: *"The atomic theory plays a part in physics similar to that of certain auxiliary concepts in mathematics; it is a mathematical model for facilitating the mental reproduction of facts"*.

It is worth stressing an important aspect, whose relevance is nowadays well appreciated in the modern study of small systems: Einstein with his study on the Brownian Motion revealed the basic role of the fluctuations, which, although very small, are important and detectable. This is a rather subtle point, and also some giants of the statistical mechanics did not understand the possibility to detect the fluctuations and their importance; e.g., Boltzmann wrote: *"In the molecular theory we assume that the laws of the phenomena found in nature do not essentially deviate from the limits that they would approach in the case of an infinite number of infinitely small molecules"*; in a similar way Gibbs remarked *"[the fluctuations] would be in general vanishing quantities, since such experience would not be wide enough to embrace the more considerable divergences from the mean values"*. Now it is clear that the fluctuations play a basic role in physics of small systems, e.g., in nano technology and biophysics.

BM is not only an important historical and technical aspect of modern physics, but it has also been the starting point for the development of the mathematical theory of stochastic processes, as well as for recent progresses in physics, biophysics, and finance. This is well exemplified by the fact that Einstein's works on BM are (still today) the most cited among his works, though the general public associates his name exclusively to relativity, and the most educated also to quantum physics.

Few years after Einstein's paper, Langevin was able to reproduce the results in a simple and elegant way by introducing the first example of a stochastic differential equation. Starting from the basic equation of the mechanics $F = ma$, Langevin split the force in two parts, a systematic (mean) one due to the friction between the colloidal particle and the fluid (i.e., the aforementioned Stokes drag) and a random one related to the collisions of the (fast) molecules of the liquid with the colloidal particle. The Langevin equation was the first non-trivial example of stochastic process. Such a mathematical field, now largely used in physics, chemistry, biology, and applied sciences, was developed in a systematic way in the 1930s by Kolmogorov with the formalization of the Fokker-Planck and Master equations for the continuous time Markov processes. A seminal work, which started from the mathematical description of the BM, is due to Wiener who introduced the idea of path integral, which stands at the basis of Feynman's formulation of quantum mechanics.

It is astonishing that, even after more than one century from the works of Einstein, Langevin, and Perrin, there are still a certain confusion and misconceptions about Brownian Motion. For instance, Popper stated that BM is a serious problem for the Second Law of the thermodynamics; and Feyerabend in his popular book *Against method* invented a Perpetuum mobile using a single molecule and explained that BM "shows" that the Second Law is wrong. Both the philosophers did not understand that the Second Law applies to macroscopic objects and not to systems made of few molecules.

To further appreciate the far reaching contribution of BM to science, it is worth noting that, despite their completely different origins, Brownian Motion and option pricing in finance are essentially described by the same equation. Few years before the paper of Einstein, the French mathematician Louis Bachelier, in his doctoral dissertation *Théorie de la Spéculation* (1900), proposed an equation for the behavior of stock's prices which is essentially the same of that describing a particle performing BM. Over the years till now, stochastic processes, in particular the stochastic differential equations, have received vast attention from the financial community and stockbrokers. For instance, the celebrated Black and Scholes theory for the option pricing is, from a mathematical point of view, nothing but an application of the Langevin equation. These deep links led, in the last decades, some thousands of people with a Ph.D. in physics or mathematics to find a job, even at high level, in banks and finance companies.

Let us conclude with few words about present times. Today technologies allow scientists to study thermal fluctuations in small systems such as colloidal suspensions. Such a possibility is very important in nano sciences. Moreover, fluctuations play a crucial role in noise-assisted transport mechanisms, also called Brownian Motors.

In conclusion, we can state that Einstein's work on the BM, which sometimes had been considered the less relevant among those of the *annus mirabilis*, with its impact on the physics of colloidal particles, other forms of soft matter (see the entry Mesoscale Systems), and of biophysical systems, had its revenge on the sub-atomic world.

Chapter 7
Chaos

Since the end of the 1980s the term *chaos* has become very pop, so to appear also in a blockbuster as Jurassic Park (1993). This was eventually due to very successful books of science popularization—as, e.g., *Chaos: making a new science* (1987) by James Gleick—the evocative power of the name "chaos" itself, which was promoted in 1975 in an influential paper by Li and Yorke, and the so-called *butterfly effect*. The latter expression appeared in 1972, when the MIT meteorologist Edward Lorenz gave a talk at the 139th meeting of the American Association for the Advancement of Science, posing the provocative question: *"Does the flap of a butterfly's wings in Brazil set off a tornado in Texas?"* As a case study for sociologists, when some scientific concept becomes popular, misconceptions and confusions are right around the corner, and *chaos* makes no exception. In this brief entry we would like to give our humble contribution to the popularization of chaos, hopefully avoiding some common misconceptions.

To start let's clarify the provocative question posed by Lorenz about 50 years ago. If you google butterfly effect, you can still find forums where people debate about the possibility that flapping wings of a butterfly (or a flock of birds) can cause a tornado. Nowadays, in forums like those you can find also people arguing the flatness of Earth, which unfortunately demonstrates that even sending men on the Moon (for duty of report, several blogs promote that no man ever landed on the Moon) or to the international space station is not enough to make people aware of the basic facts of science. Back to the butterfly wings, as Lorenz himself clarified in his speech *"whether, for example, two particular weather situations differing by as little as the immediate influence of a single butterfly will generally after sufficient time evolve into two situations differing by as much as the presence of a tornado. In more technical language, is the behavior of the atmosphere unstable with respect to perturbations of small amplitude?"* So the point is about the stability of the atmosphere to tiny disturbances, and more in general if we have a deterministic phenomenon (see the entry Determinism for a broader and historical view), e.g., which can be described in terms of differential equations, how precisely we need to specify its initial state

© Springer Nature Switzerland AG 2021
M. Cencini et al., *A Random Walk in Physics*,
https://doi.org/10.1007/978-3-030-72531-0_7

in order to be able to predict its future evolution. This question does not apply only to very complicated systems such as the atmosphere but appertains to all systems that admit a description in terms of differential equations, including the dynamics of celestial bodies (as in the Solar system) that follow Newton's law of gravitation,[1] as discovered at the end of the nineteenth century by Poincaré (see below and the entry Poincaré).

In 1963, Lorenz, while studying by means of computer assisted calculations a very crude model of the atmosphere, discovered that the above question has a non-trivial answer. The Lorenz model, obtained as a simplification of the problem of convection (roughly the motion of air heated from below and cooled from above), consists of only three variables X, Y, and Z representing the amplitude of convective motion, the temperature difference between ascending and descending fluid parcels, and the deviation from a linear temperature profile. The evolution of these three variables obeys the following set of ordinary differential equations:

$$\frac{dX}{dt} = \sigma(Y - X)$$
$$\frac{dY}{dt} = rX - XZ - Y$$
$$\frac{dZ}{dt} = XY - bZ,$$

where the constants σ, r, b are three parameters related to the physics problem: σ is the so-called Prandtl number that is the ratio between fluid viscosity and thermal diffusivity; r the Rayleigh number related to the temperature difference between top and bottom—the main control parameter; and b relates to the geometry of the system. In his original work, Lorenz fixed $\sigma = 10$, $r = 28$ and $b = 8/3$. The above system of equations is non-linear[2] and cannot be solved with the only help of a pen and a paper, that is why Lorenz used a digital computer to solve it. The history of the discovery of chaos, though the name came later, in the above system is a bit anecdotal. Lorenz himself said in his book, *The essence of chaos*, that while computing the time evolution of the three variables and printing the results, he found that trajectory's character was very irregular (i.e., non-periodic, as clear from Fig. 7.1a displaying the behavior of the Lorenz model). In order to better understand, he repeated the computation and initialized the values of the variables cutting some digit, say that the value of X was 0.506127 and he entered 0.506. Then he discovered that very soon *"the numbers being printed were nothing like the old ones"*. After excluding possible sources of error, he understood that tiny differences in the initial conditions were amplified in the course of time. Therefore, not only the trajectories generated by the (apparently simple) Lorenz equations were highly irregular but they also were

---

[1]Which, by the way, is an approximation of Einstein general relativity, but for our purposes here we do not need to enter this difference.

[2]Non-linearity means, for instance, that in the right-hand side of the above equations there appears polynomials of degree larger than 1 in the variables, e.g., XY or XZ.

highly unstable, indeed the difference between the trajectories with slightly different initial conditions amplified exponentially in time (see Fig. 7.1b). Strangely enough, on the other hand, very different initial conditions when evolved are *attracted* by a double wings structure as depicted in the figure displaying the Lorenz attractor, denoting a certain degree of *stability* (as shown in Fig. 7.2).

The above phenomenology was strongly at odds with the kind of behaviors scientists were used to at that time. Usually stability was associated with fixed points, namely, with reference to the equations of the Lorenz model, those values of X, Y, and Z such that the right-hand sides are simultaneously equal to zero. Fixed points can be stable in that small deviations are damped in time so that at long times the trajectories set on the fixed point values, or unstable when small perturbations are amplified. Think, for example, to a rigid pendulum subject to air friction. When the massive bob is exactly above the pivot point a very small deviation to the left or right

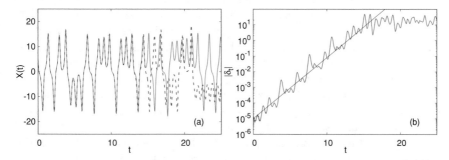

Fig. 7.1 Behavior of the Lorenz model. Panel **a** shows the evolution of the X variable for two initial conditions differing only on the fifth digit after the dot. Note the irregular behavior and the fact that soon the two trajectories go far away from each other. Panel **b** displays the time evolution of the distance, δ_t, between the two trajectories (red curve) in logarithmic scale on the vertical axis. The black solid line shows the exponential behavior $\delta_0 \exp(\lambda t)$ with Lyapunov exponent $\lambda \approx 0.905$

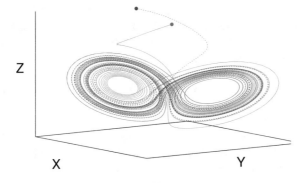

Fig. 7.2 The *strange attractor* of the Lorenz model: The two colors code the evolution of the three variables X, Y, Z starting from two very different initial conditions the bullet points. After a short time, the two trajectories while remaining distinct and never intercepting at the same time moment, evolve in the same "structure", a subset of the three dimensional space

leads the bob to fall left or right (unstable fixed point), on the other hand at long times any initial condition ends with the bob steady below the pivot point (stable fixed point). Actually, this is a severe simplification of the possible form of stability. For instance, the possibility of so-called limit cycles, i.e., stable periodic trajectories, was also known since the 1920s thanks to the studies of non-linear oscillators by Andronov, van der Pol, and Wiener and of non-linear equations describing the dynamics of populations due to Lotka and Volterra. However, at the time of Lorenz, nobody had ever seen a "stable object" like that the double wings structure depicted in Fig. 7.2. We had to wait 1971 for Ruelle and Takens to name it as *strange attractor*, another evocative name which likely contributed to make chaos so pop.

Nowadays, actually since the 70s and a lot more since the 80s, we know that the phenomenology discovered by Lorenz is quite generic for non-linear dissipative dynamical systems. The term dissipative has to do with the fact that if we start with a set of initial conditions (starting values for the system variables) occupying a certain volume, such volume is not preserved by the dynamics. To make it clearer, assume that at time $t = 0$ we take as possible initial conditions for the Lorenz model all the values of X, Y, Z belonging to a cub-let of given side, and evolve each initial condition with the Lorenz differential equations, then at long times they occupy a portion of zero volume, indeed the strange attractor above depicted is closer to a sheet than to a cube. Specifically, it is a fractal object with dimension $D_F \approx 2.05$ (see the entry Fractals). This is not a peculiarity or oddity of the Lorenz model, chaotic dissipative systems are characterized by strange attractors with fractal geometry. Non-chaotic dissipative systems also have attractors, these can be either stable fixed points (here the dissipative nature is even clearer, all the initial conditions in the basin of attraction asymptotically end in the fixed point which is zero dimensional, think again to the pendulum with friction), or stable limit cycles as mentioned above. What makes chaos intriguing and interesting is that this form of *strange* stability is accompanied by the instability of trajectories discovered by Lorenz: loosely speaking it is like if almost every point of the system behaves as the unstable fixed point of the pendulum with friction.

The phenomenology of chaos is not limited to ordinary differential equations but can be found also in discrete-time dynamical systems, or maps. These kinds of systems can be seen either as discretization of a continuous time system via a technique due to Poincaré, namely, one consider the intersections of the trajectory with a given set (e.g., a plane) or as systems with an intrinsically time discrete nature. As an example of the latter, you can think of a population described in terms of generations. A popular example is the logistic map $x_{t+1} = r x_t (1 - x_t)$, where x_t represents the normalized abundance of a species which reproduces (between generation t and $t + 1$) with a rate r and, due to resource limitation, competes with conspecific with a death rate proportional to the population abundance squared. Changing the value of the reproduction rate r, the logistic map displays a variety of dynamics from fixed points to periodic orbits to chaos. Another nice example is the Hénon map, which can be seen as a map of the plane in itself defined by the following two equations:

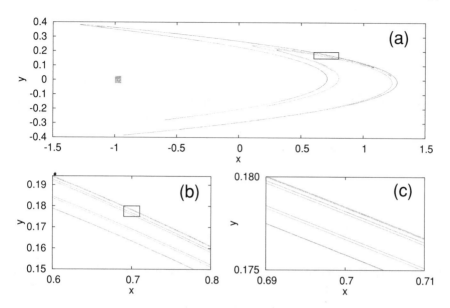

Fig. 7.3 The *strange attractor* of Hénon map. In panel **a**, the blue square represents a spot of 10^4 initial conditions, the purple dot display the evolute of that spot after 10^3 iterates of the Hénon map with $a = 1.4$ and $b = 0.3$. Panel **b** displays a blow-up of the rectangle region in (**a**) and panel **c** a blow-up of the rectangle region in (**b**)

$$x_{t+1} = 1 - ax_t^2 + y_t \quad \text{and} \quad y_{t+1} = bx_t,$$

with a and b two constants. As shown in Fig. 7.3a we see that an initial spot of initial conditions evolve into a strange attractor with fractal features, indeed the Hénon map is an example of discrete-time dissipative dynamics. To make apparent one of the most well-known properties of fractal objects in Fig. 7.3b we show an enlargement of a small portion of the attractor, which is again enlarged in Fig. 7.3c: at each blow-up a series of stripes appear which self-similarly resemble themselves on finer and finer scale.

The instability of trajectories of chaotic systems can be characterized in three (connected) ways. The more straightforward way is based on the observation that, as mentioned above, tiny differences in the initial conditions, or if you prefer an incomplete knowledge of the initial state of the system, is exponentially amplified in the course of time. It is then natural to characterize it in terms of the exponential rate of amplification of such difference. This leads to the introduction of the Lyapunov exponent, λ. Let x_0 be the initial state of dynamical system, being it an ordinary differential equation or a discrete-time map. Now assume a very close (mathematically speaking infinitesimally close) state $x_0' = x_0 + \delta_0$. Then evolve the two states according to the given dynamics, if this is chaotic one has that at long times (mathematically in the limit $t \to \infty$) the difference between the two states will increase exponentially with rate λ, i.e., $|\delta_t| \approx |\delta_0| \exp(\lambda t)$, where the mathematically correct definition of λ is given by[3]:

$$\lambda = \lim_{t \to \infty} \lim_{|\delta_0| \to 0} \frac{1}{t} \ln \frac{|\delta_t|}{|\delta_0|},$$

i.e., one should first consider infinitesimally small perturbations of the initial state and then wait a lot of time in order to see the proper value of the Lyapunov exponent. The exponential growth of the uncertainty on the initial state is basically the essence of the *butterfly effect* meant by Lorenz and is often referred to as sensitive dependence on initial conditions. A clear consequence of it, which is often (to some extent incorrectly) ascribed as the reason for which we cannot forecast the weather, is that if we fix a certain level of tolerance, say Δ on the initial state we can only make prediction up to a time

$$T_p = \frac{1}{\lambda} \ln \frac{\Delta}{\delta_0}.$$

The ln in the above formula makes very hard to improve the prediction time T_p: even if the error δ_0 is reduced by a factor 10, the time T_p improves only by a factor slightly larger than 2. As discussed in the entry Prediction, while it is in general true that having a positive Lyapunov exponent makes long-term predictions impossible, it is not the end of the story when confronting with complex issues as whether forecasting.

Before discussing other aspects, it is worth specifying that the above definition of λ refers only to the maximal Lyapunov exponents, for systems involving many variables, say N, we can define N Lyapunov exponents describing how an infinitesimal spot of initial conditions is deformed. In general, one positive exponent is enough for having chaos but, of course, there can be many positive Lyapunov exponents, though the sum of all N Lyapunov exponents, if the system is dissipative should be negative. From a conceptual point of view, the importance of the Lyapunov exponents relies on a deep mathematical property: they are topologically invariant. With this obscure term, mathematicians want to mean that they represent an intrinsic property of the system, i.e., if we make an invertible change of variables and change accordingly the equation of motions, the values of the Lyapunov exponents remain unchanged.

Another way to characterize the instability of a chaotic system is to considering it as a source of information and to adopt an information theory point of view. This approach, detailed in the entry Information Theory, naturally leads to the concept of Kolmogorov-Sinai entropy (see also the entry Kolmogorov). Finally a third way to approach the characterization of chaos is that of algorithmic complexity. In this case we can interpret the evolution laws of a system (as, e.g., the Lorenz model) as an

---

[3]Notice that, in principle, λ may depend on the initial state x_0, however, typically in dissipative systems λ is unique and the same for almost all initial conditions. The "almost all" has a specific mathematical meaning which to be fully understood would require introducing some notion of measure theory. To simplify things we can say (a bit imprecisely) that randomly choosing the initial condition the probability that the limit defined in the equation assumes the value λ is 1. This property has another evocative, slightly obscure name, that is *ergodicity*, when it holds the system is said to be ergodic (see the entry Statistical Mechanics for details about the concept of ergodicity).

algorithm to produce a given sequence, the values of the system variables at a given time. The presence of chaos manifests in the fact that the vast majority of sequences originating from a chaotic system are "complex" (see entry Complexity Measures, for details).

Up to now we have drastically simplified Lorenz's findings and somehow cut short a longer history that actually starts much earlier, namely, in 1887 when Poincaré won the mathematical competition, in honor of King Oscar II of Sweden, for the 3-body gravitational problem (see the entry Poincaré). Exactly 200 year before, in 1667, Newton solved the two-body problem, showing that the three Kepler's empirical laws could be derived solving the gravitational equations introduced by himself, which thanks to symmetries can be reduced to a set of two differential equations. Since then, notwithstanding the progresses of celestial mechanics, nobody could solve a small perturbation of the original two-body problem, i.e., when a third mass (no matter how small) enter the game. In the attempt to do this, Poincaré actually demonstrated that such solution does not exist in general. Even if the reason found a name—chaos—only much later, Poincaré deeply understood already at that time the nature of chaos. The best summary for his understanding is repeating his words written in 1903 (60 years before Lorenz's works): "A very small cause that escapes our notice determines a considerable effect that we cannot fail to see, and then we say that the effect is due to chance. If we knew exactly the laws of nature and the situation of the universe at the initial moment, we could predict exactly the situation of that same universe at a succeeding moment. But even if it were the case that the natural laws had no longer any secret for us, we could still only know the initial situation approximately. If that enabled us to predict the succeeding situation with the same approximation, that is all we require, and we should say that the phenomenon had been predicted, that it is governed by laws. But it is not always so; it may happen that small differences in the initial conditions produce very great ones in the final phenomena. A small error in the former will produce an enormous error in the latter. Prediction becomes impossible, and we have the fortuitous phenomenon".

It should be noticed that, differently from Lorenz model, the gravitational equations constitute a conservative dynamical systems. With reference to the above explanation of dissipative, for conservative systems volumes are preserved. This means, for example, that there cannot exist attractors of any type. This mathematical difference with dissipative systems has, of course, important consequences in the manifestation of chaos and the way chaos sets in, but this goes beyond the scope of this short non-technical discussion.

After more than 100 years, Poincaré words are crystal clear in explaining the essence of chaos and are absolutely not outdated. In particular, they highlight a crucial aspect of chaos and more in general of deterministic laws (see the entry Determinism). Even if we are able to capture the "true" deterministic laws of the universe (assuming this is possible) our limitation as humans in knowing the actual state of the universe (i.e., with infinite digits of precision) will impede our possibility to predict the future. Rephrasing in a motto we can state that through chaos there is room for chance also in a deterministic universe. Again a quite evocative way of speaking about chaos which, we think, should not be taken too literally to avoid misconceptions. However, it is clear that thanks to the irregularities of chaotic evolutions (see again Fig. 7.1a) and to the instability to perturbations, the use of probabilistic concepts (typically used in characterizing phenomena of chance) in chaotic systems is basically mandatory, again a manifestation of our (human) limitations.

Chaos over the last three decades has been promoted to a paradigm shift for many aspects, and used/abused to explain many things. This is not the proper forum to explain all the abuses and misconceptions, the interested reader can find a couple of references on these aspects at the end of this book. However, we like to conclude by stressing that the discover of chaos contributed in blurring several conceptual opposi-tions, such as order/disorder, random/non-random, stable/unstable, simple/complex, showing that irregular and complex behaviors can emerge from well formulated (seemingly) simple dynamical laws.

Chapter 8
Complexity Measures

The words *complex* and *complexity* are nowadays commonly used in many branches of science, and not only. Often we can read in popular or technical articles the terms "Complexity theory", "Complexity Science" or "Science of Complexity", which do suggest the existence of a well defined research subject with its own concepts and tools. This is actually not true in a broad sense. A thoughtful discussion on complexity in general and on its use and misuse in science would be very interesting but it would easily become too technical. For this reason, here, we shall confine the subject. In particular, we shall focus on *Algorithmic Complexity* (sometimes dubbed with the names of its inventors Kolmogorov-Solomonoff-Chaitin Complexity or Algorithmic information theory) of one-dimensional sequences (strings) of numbers. This notion, besides its interest per se, allows for reconsidering other entries of this book (e.g., Information Theory, Chaos and Laws, levels of description and models) from a different viewpoint and thus (we hope) deepening their understanding.

In order to illustrate the basic idea of the approach devised by Kolmogorov and, independently, by Solomonoff and Chaitin, it is useful to consider a simple example. Consider playing a coin toss game and to record the series of head/tails (1/0) in N repetitions of the coin flipping. If we assume the coin to be fair (unbiased) at each flip the probability to obtain 0 or 1 is 1/2 and, since each flipping is independent from the others, the probability to observe 00, 01, 10, or 11 in two consecutive flipping is 1/4. In general, N consecutive games will generate a sequence of 0s and 1s with probability $1/2^N$. For instance, two realizations of $N = 30$ coin flipping repetitions can generate the two following sequences:

$$S_1 = 101101011010100100100101100001$$
$$S_2 = 001001001001001001001001001001,$$

with the same probability $p = 1/2^{30}$. As discussed in the entry Information Theory (see also Entropy), if we are asked to quantify the information contained in the two

© Springer Nature Switzerland AG 2021
M. Cencini et al., *A Random Walk in Physics*,
https://doi.org/10.1007/978-3-030-72531-0_8

above sequences, our answer would be that their information content is the same and it is equal to $\log_2(1/p) = 30$ bits. While this answer leaves us with the satisfaction of having correctly employed the Shannon definition of information, it negates our intuition. Indeed the second sequence looks much simpler (and regular) than the first one. For instance, to reproduce (communicate) it would suffice to say:

PRINT '001' FOR 10 times,

i.e., 3 bits to specify 001 (which is repeated) and 4 bits[1] to express the number of repetitions plus some additional bits for the instruction PRINT and FOR. On the other hand, it appears that the most concise description of the first sequence should look like

PRINT '101101011010100100100101100001',

which uses 30 bits, i.e., the sequence itself, plus some other bits for the PRINT. In other terms our intuition would call the first sequence more complex that the second one, as it is harder to describe. Here harder clearly means that we need more bits to specify it than the other one.[2] This intuitive notion of complexity is well expressed by the definition "a complex object is an arrangement of parts, so intricate as to be hard to understand or deal with", according to Webster Dictionary.

Why is the answer of information theory at odds with our intuition about the complexity of the two above sequences? If we consider the whole ensemble of sequences we can obtain flipping 30 times the coin, then claiming that the information content is on average 30 bits is completely correct and agrees with our intuitive expectation. However, in the above example we implicitly assumed a different perspective: we are not asking about the information content of an ensemble of sequences but we are interested in those two "specific" instances. In other words the entropy of information theory is characterizing the source (i.e., the whole possible messages emitted by the source, here the coin tossing) and not the single message.[3] Hence, e.g., for the practical problem of compressing a message (and then communicating it) information theory is fine for the generic messages emitted by the source but can be not adequate for the single message. The change of perspective here is that we are interested in defining the complexity (information content) of a specific message regardless of the source (i.e., the probabilistic scheme) which generated it. How can we define the information content of a single message, i.e., its complexity?

Algorithmic complexity is precisely answering this last question, and the approach it follows is basically a formalization of the way we discussed the difference between the two sequences above. The two descriptions for the sequences S_1 and S_2 are

---

[1] The number 10 can be expressed as $10 = 0 \cdot 2^0 + 1 \cdot 2^1 + 0 \cdot 2^2 + 1 \cdot 2^3$, thus it is codified by the four bits 0101.

[2] For $N = 30$, the difference in bits of the two descriptions might be not astonishingly large, but it will become so when considering similar sequences for large N.

[3] In the entry Information Theory, we mentioned the theorem of Shannon-McMillan or of the asymptotic equiprobability, which distinguishes typical and non-typical sequences. In this perspective, clearly the sequence S_2 is non-typical.

essentially two programs[4] which can be executed on a computer to reproduce the two sequences. The algorithmic complexity, or the information content, of a sequence S can be defined as the length in bits of the shortest program, $\mathcal{P}_\mathcal{M}(S)$, which run on a computer, $\mathcal{M}$, prints the sequence S. In order to make the above definition meaningful, however, we need two important specifications.

First, which computer shall we use? In principle, we can devise a computer that, upon receiving as input 0 prints the sequence S_1, and in which all other programs starts with 1. In this case the more complex sequence would result having lower complexity. To avoid this arbitrariness one should use a universal computer $\mathcal{U}$ (a notion due to Alan Turing), which can simulate any other computer using a finite emulator program. In this way the algorithmic complexity would be defined except for a constant (the length of the emulator program).

Second, we should require that the program stops after printing the sequence S. This is a delicate point which entails the issue of the so-called halting problem (implied by Turing's theorem of uncomputability), namely, the impossibility to determine whether an arbitrary program will stop or run forever. Moreover, the request of stopping after printing the sequence is crucial to define the mutual algorithmic complexity between two (concatenated) sequences and to define a measure of the probability of the sequence even if we do not know the probabilistic scheme which generated it. While all these aspects are interesting and very deep, we will not consider them here for space constraints and to maintain the discussion not too technical. The interested reader can deepen such aspects consulting the dedicated literature (see also the section Further reading at the end of this book).

In summary, the algorithmic complexity, $C_\mathcal{U}(S)$, of a sequence S with respect to a universal computer $\mathcal{U}$ is defined as

$$C_\mathcal{U}(S) = \min_{\mathcal{P}_\mathcal{U}(S) \text{stopping after printing } S} Length[\mathcal{P}_\mathcal{U}(S)],$$

which, apart from a constant, is independent of the computer as we use a universal one. The fact that such constant is inessential can be understood by considering the algorithmic complexity per symbol that, given a sequence S of N symbols (zeros and ones), is defined as

$$c_\mathcal{U}(S) = \lim_{N \to \infty} \frac{C_\mathcal{U}(S)}{N},$$

so that all constant contributions divided by N for N large disappear. Complex sequences will have $c_\mathcal{U}(S) > 0$ and non complex ones $c_\mathcal{U}(S) = 0$. For instance, consider the number $\pi = 3.14159265358979323844\ldots$, is it algorithmically complex? It is an irrational number, any statistical test of randomness would suggest that the number is "complex" as the numbers $[0, 9]$ and all the couples, triples, etc. essentially

---

[4]A program in machine language is essentially a sequence of zeros and ones which encodes input and instructions to obtain an output, i.e., the result (see also the entry Computer, algorithms and simulations).

occur with equal probability. Considering the infinite sequence of digits of π and computing the entropy per symbol would give thus a finite entropy. However, simple programs (e.g., based on polygon approximation to a circle or series representation of π) can print an arbitrary number, N, of digits using only about $\log_2 N$ bits, so that the algorithmic complexity per symbol of π is actually zero. Indeed π is not just an irrational number but it is a very specific one. Similar considerations apply to other irrational numbers as $\sqrt{2}$, the Euler number $e = 2.7182818284\ldots$ and many other numbers of this kind.

The above example and the previous discussion may lead the reader to consider entropy (of information theory) and algorithmic complexity to be completely unrelated. This is not the case. Given an ensemble of sequences of N (with N very large) symbols emitted by a source, one can compute the algorithmic complexity of each sequence and average their values according to the probability of occurrence of each sequence. It is possible to show that such average algorithmic complexity is (asymptotically for $N \to \infty$) equal to the Shannon entropy of the source, and that most of the sequences have algorithmic complexity per symbol equal to the Shannon entropy of the source. These results stem from the Shannon-McMillan theorem which we briefly discussed in the entry Information Theory. Algorithmic complexity, however, is somehow more fundamental than information theory as it defines information by requiring no other knowledge than the individual messages. From the point of view of physics, however, the use of statistical quantities such as entropy is typically more valuable as one is not interested to understand or characterize a single phenomenon or pattern but the ensemble of possible ones emerging from a specific system.

It is interesting now to reconsider the Bernoulli shift map, a simple example of chaotic system, that we used in the entry Information Theory to show how entropy can be used to characterize chaotic systems and that entropy is indeed linked to the Lyapunov exponent, which quantifies the sensitive dependence on initial conditions—the hallmark of chaotic systems (see the entry Chaos). Intuitively, we should expect sequences obtained by chaotic systems to be algorithmically complex. On the other hand, this expectation seems to contrast with the fact that the sequences generated by the Bernoulli shift map can be obtained by means of a very short program, schematically described as follows. Take as initial condition x_0 a number between 0 and 1. Knowing the number x_t at time t generate the next one with the rule $x_{t+1} = 2x_t$ mod 1. The symbol s_t at time t is $s_t = 0$ if $x_t < 1/2$ and $s_t = 1$ if $x_t \geq 1/2$. Print s_t for $t = 1, \ldots, N$, with N as large as desired, and stop. This seems to be a very concise description (program) to generate the sequence $S = s_0 s_1 \ldots s_N$, which would naively suggest the algorithmic complexity per symbol to be 0. Where is the trick? In order to understand this point, it is useful to recall the binary description of a real number between 0 and 1, for instance, the initial condition can be written as $x_0 = \sum_{k=0}^{\infty} s_k 2^{-k-1}$, where $s_k = 0$ or 1. Notice that the use of s_k to denote the coefficient of the binary expansion is intentionally the same as that used for the symbolic sequence generated by the map: if $s_0 = 0$ (or 1) clearly $x_1 < 1/2$ (or $\geq 1/2$). Then applying the rule $x_{t+1} = 2x_t$ mod 1, one can easily realize that $x_1 = \sum_{k=0}^{\infty} s_{k+1} 2^{-k-1}$ $\ldots x_t = \sum_{k=0}^{\infty} s_{k+t} 2^{-k-1}$. In other terms to generate that specific sequence of N symbols $s_0 \ldots s_N$ we need to prepare the initial condition exactly using that sequence.

Therefore, the algorithmic complexity of the sequence generated by the Bernoulli shift map is essentially the algorithmic complexity of the binary representation of the initial condition. Martin-Löf (who was a student of Kolmogorov) proved that almost all real numbers in the interval $[0, 1]$ admit a binary representation which is algorithmically complex.[5] Therefore, in the Bernoulli map, the algorithmic complexity of the derived sequences is due to the number of bits needed to specify the initial condition, i.e., we are basically in the situation of the sequence S_2 in our first example: the sequence has non-zero algorithmic complexity per symbol.

The Bernoulli shift map example, though very simplified, shows that in chaotic systems the algorithmic complexity is tightly linked to the problem of specifying the initial condition and to the sensitive dependence on initial conditions, i.e., the Lyapunov exponent. Shortly, since a small difference on the initial condition grows exponentially with the Lyapunov exponent, $\delta x_t = \delta x_0 \exp(\lambda t)$, if to generate the symbolic sequence of length T we use a resolution ϵ, in order to reproduce a specific sequence we have to require $\delta x_T \leq \epsilon$. Consequently, the input must be specified using at least $T\lambda / \ln 2 + \log_2(1/\epsilon)$ bits, i.e., a number of bits growing linearly with T, so that the algorithmic complexity per symbol will be essentially $\lambda / \ln 2$.

There is, however, a very subtle and deep issue with algorithmic complexity. By definition it requires to determine the shortest program able to print the sequence of interest, the problem is that we can never be sure to have found the shortest one. Said in technical terms the algorithmic complexity is not computable. This apparently odd fact is deeply linked to a famous result of mathematical logic, namely, Gödel's incompleteness theorems, and the aforementioned halting problem. A detailed discussion of this profound issue goes far beyond the scope of this short survey, so we limit to provide a glimpse of the problems underlying the computability of algorithmic complexity mentioning the famous Berry's paradox *"Let N be the smallest positive integer that cannot be defined in fewer than twenty English words"*. As one can verify, such a statement defines the number using only 17 words! Contradictory statements of this kind can be used to prove the uncomputability of algorithmic complexity.

In spite of its uncomputability, algorithmic complexity allows for founding information theory starting from messages (this was probably the driving goal of Kolmogorov, see the related entry), ignoring the source, and it also provides hints on how to approach and solve very practical problems such as the compression of strings of data. Actually, algorithms for the compression of strings of data can be used to (partially and practically) overcome the problem of the uncomputability of algorithmic complexity. For instance, popular compression algorithms, such as gzip and zip, provide a way to encode sequences of numbers into compressed versions with a particular encoding. The length of such compressed version of the original sequences can be used as an upper bound to the true (uncomputable) algorithmic complexity. Clearly, this method will fail to recognize as non-complex numbers like π, but will provide reasonable estimations for generic sequences. Given their everyday usage it

---

[5]Only countable many are not algorithmically complex, these are all the rational numbers that, indeed, under the shift map give rise to periodic sequences.

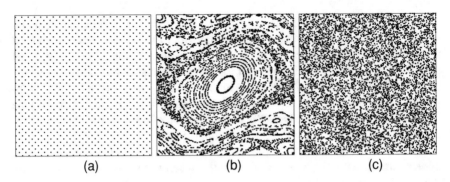

Fig. 8.1 Three illustrative examples patterns to test your intuitive notion of complexity (after Grassberger)

is worth in the following briefly describing the way these algorithms work, in particular we will focus on one of them based on the ideas developed by Ziv and Lempel. Given a sequence of symbols S, one can segment it in words $W_0 W_1$, $W_2 \ldots$ such that $W_0 = \emptyset$ (i.e., the empty word), and W_{n+1} is the shortest new word following W_n. As a practical example consider the sequence $S = 101101011010100100100101011000$. We can break it as $S = (1)(0)(11)(01)(011)(010)(10)(0100)(100)(101)(1000)$. In this way, word W_n is equal to a previous word W_k with $k < n$ plus one single symbol s_{new}. Then the words are encoded as the pairs (j, s_{new}). So the compressed version of S will look as $S = (0, 1)(0, 0)(1, 1)(2, 1)(4, 1)(4, 0)(1, 0)(6, 0)(7, 0)(7, 1)(9, 0)$. While for this short sequence it does not seem we compressed much, when words start to be long this trick compresses them a lot exploiting previously seen (shorter words).

While the uncomputability of algorithmic complexity may seem a severe limitation, it does not undermine its conceptual importance. However, algorithmic complexity is often criticized as a measure of complexity because, while it surely provides a measure of randomness, it is unclear whether this corresponds to our intuition of complexity. To test your opinion we ask you to look at the patterns in Fig. 8.1.

Which pattern between (a), (b), and (c) do you consider more complex? We are sure nobody would call (a) as complex. You will probably be in doubt between (b) and (c), likely most of you will consider (b) the most complex one, especially if you are told that (c) was obtained by simply drawing random dots.[6] Well, pattern (c) has actually a larger algorithmic complexity just because is more random. Such a simple example shows that our intuitive notion of complexity does not correspond to randomness. Likely the pattern (b) seems to us more complex as we see some order and correlations together with some disorder, it is more sophisticated that (c) and (a). Can we quantify this? In the last decades many attempts have been made, you can find dozens of complexity measures. Their multitude is a manifestation of how complex

---

[6]Googling "complex pattern" which basically means checking typical patterns considered complex by people, you will likely confirm that (b) is the most complex pattern.

is complexity,[7] and it is due to the impossibility to define complexity on a purely objective basis disregarding the role of the observer perceiving the complexity.

We mention here just one of these attempts originally proposed by Charles H. Bennett: the *logical depth* that extends algorithmic complexity in an interesting way. According to Bennett the logical depth of a sequence is related to the time needed by a universal computer to actually run the shortest program able to generate the sequence. You can see that this is a sound measure of complexity thinking about the DNA. We know that DNA codes for our genes and can be seen as the program for assembling an organism (indubitably a complex object). The DNA is a long sequence of symbols (from the four letters alphabet A,C,G, and T), 3 billion of them make the human genome. It is a very redundant sequence, clearly non-random and so with lower algorithmic complexity than an equivalent length sequence obtained by randomly drawing the four symbols. It likely started from a very short sequence and it becomes what it is running a short program based on random mutations and selection which, however, has run for a very long time (the evolution) to generate the complexity of our body. Clearly, the DNA must have a large logical depth.

We conclude mentioning that both algorithmic complexity and logical depth can be seen as a formalization of Occam's razor according to which *"Entia non sunt multiplicanda praeter necessitatem"*, i.e., more things should not be used than are necessary when explaining something (the shortest program). Although using these few principles to explain a (complex) phenomenon may require a long time (the logical depth). Reconsidering the entry Laws, levels of description and models, we can ultimately think that a theory of complexity, if it exists, should be actually a theory of modeling.

---

[7]Forgive us for the tautology.

Chapter 9
Computers, Algorithms, and Simulations

A computer is an electronic device that performs calculations and operations based on precise instructions. One thinks of the computer as a more or less bulky device with keyboard and video attached, but in reality any device capable of performing pre-established operations in an automatic way can be called a computer. From the seventeenth century up to some years after the end of World War II the term "computer" was used to designate humans who compute, i.e., people performing mathematical calculations.[1] For instance, Alan Turing, whose key role on the development of the electronic computer will be examined below, in 1950 describes the digital computer referring to the human one *"The idea behind digital computers may be explained by saying that these machines are intended to carry out any operations which could be done by a human-computer. The human-computer is supposed to be following fixed rules; he has no authority to deviate from them in any detail"*. For instance, the film *Hidden Figures* recently celebrated the figures of (mainly black) women employed as computers that, at the beginning of 1960s, played a key role in the space race between USA and URSS. The film also well describes the transition, happened in those years, from human to electronic computers.

Nowadays (electronic) computers are almost everywhere, from satellite navigators to food processors, from smart watches to cash machines. Without computers modern life would not be the same and, similarly, even modern science would undergo significant changes. Each computer can be programmed to perform very different tasks, from very practical ones, e.g., fill out tax declaration, to the most theoretical ones, e.g., retrieve big prime numbers. Before coding a specific task in a computer language, this task has to be precisely defined in what it is called *algorithm*, namely, the aforementioned *"fixed rules"*.

The term algorithm derives from the Persian mathematician Muḥammad ibn Mūsā al-Khwārizmī (c. 780–c. 850), who has been a central figure along the path leading

---

[1] See also the entry Richardson for a discussion on human computers.

© Springer Nature Switzerland AG 2021
M. Cencini et al., *A Random Walk in Physics*,
https://doi.org/10.1007/978-3-030-72531-0_9

to modern mathematics. He studied and worked at the Bayt al-Ḥikma (the House of Wisdom) in Baghdad, one of the greatest cultural institutions of all times that had an important role, in particular, for in the translation into Arabic of all the scientific texts known at that time and written in other languages. One of the most important contributions to mathematics due to al-Khwārizmī has been *Algoritmi de numero Indorum* (On the Indian Numerical Calculation, whose Arabic version has been lost and it is known only through a Latin version of the thirteenth century) in which he has given a complete exposition of the Indian numbering system from which the modern numbering system is derived.[2] Another of his fundamental works has been *al-Kitāb al-mukhtaṣar fīhisāb al-jabr wa l-muqābala* (Short book on the calculation of completion and reduction, and it is from the term "al-jabr" in the Arabic title of the book that the modern term Algebra came from) in which al-Khwārizmī presented first- and second-order algebraic equations with detailed step-by-step procedures (and the modern term algorithm denotes procedures of this type) to solve them.

A witty definition of algorithm: "A word used by programmers when they do not want to explain what they did" is quite famous in the computer scientist's community, but actually it is easy to define what an algorithm is. An algorithm is something very similar to a recipe: it tells you how to accomplish a task by performing a number of fixed and well defined steps. However, unlike a recipe, each step of an algorithm must be specified with absolute precision. Indeed to define an algorithm there is no room for statements like "Add a pinch of salt" or "Mix for a few minutes": the terms "pinch" or "few minutes" are too vague (what is few? ten, five?). An algorithm must unambiguously specify all those logical steps whose execution inevitably leads to the solution of a certain problem. Usually, for its execution, an algorithm needs a defined set of inputs and, as a result, produces a defined set of outputs. For example, the algorithm used to find the smallest of ten numbers needs in input ten numbers and produces as output one number (hopefully the smallest of those in input).

The introduction of algorithms in mathematics goes back thousands of years, long before the work of al-Khwārizmī. There are plenty of examples about ancient algorithms introduced to solve specific problems, from the Babylonian-Sumer method of extracting roots (that is one of the first documented examples of mathematical algorithms) to Euclid's algorithm to compute the greatest common divisor or the sieve of Eratosthenes for finding prime numbers. During the history of science, several scientists developed smart algorithms to rapidly perform calculations. For instance, the Newton-Raphson method (1669) to obtain accurate values of some quantities by computing the zeros of real valued functions or the Fast Fourier Transform (FFT) introduced by Gauss, in an unpublished work of 1805, to interpolate the orbit of asteroids Pallas and Juno with only few observations. Now FFT has an instrumental role in many technological applications and medical imaging, e.g., in the Computed Axial Tomography (or CAT scan).

---

[2]For instance, the "zero" was not present in ancient Greek mathematics and was taken from Hindu numerals by al-Khwārizmī, who called it *sifr* ("empty"). The Hindu-Arabic numerals with the zero were spread in Europe only at the beginning of thirteenth century mainly under the influence of the Italian mathematician Fibonacci, with his *Liber Abaci*, who read latin translations of al-Khwārizmī.

It is worth mentioning that in the first half of the nineteenth century there was a first attempt to introduce a completely automatic mechanical computer, developed by Charles Babbage (1791–1871) with his Analytical Engine described in the work *On the Mathematical Power of the Calculating Engine* (1837). Ada Lovelace (1815–1852, mathematician and writer, daughter of Lord Byron) worked on the project and must certainly be remembered as the first computer programmer, as she developed an algorithm to generate Bernoulli numbers, specifically intended to be processed by Babbage's machine. The British government, while considering Babbage work exceptional, a general purpose computer whose structure was essentially similar to those of the electronic era, decided not to finance it. The first electromechanical computer, Z3, was built in 1941, more than a 100 years after Babbage's visionary work.

When the distinction between scientific disciplines was not as rigid as now, and scientists dealt with very different subjects, numerical methods and algorithms were perfectly accepted as a mean to obtain specific results. But starting from the mid-1800s, the differentiation between disciplines and the specialization became a distinctive aspect of scientific knowledge, and numerical methods and algorithms began to be considered, especially among pure mathematicians, marginal and uninteresting. It took a while to accept these methodologies as an integral part of the scientific knowledge. For instance, only in 1932, Mauro Picone[3] held one of the first courses on numerical methods ever offered at an Italian university, "Calcoli Numerici e Grafici" ("Numerical and graphical calculation"). However the course was given in the School of Statistical and Actuarial Sciences, because Picone's colleagues in the Mathematical Institute at the University of Rome judged the course not having the necessary characteristics to be included in the Master degree in Mathematics. The importance of algorithms also for very theoretical issues such as the definition of randomness or the foundation of Information Theory is now well recognized, as discussed in the entry Complexity Measures.

Only after the development of electronic computers the importance of algorithms and numerical methods was recognized by the majority of the scientific community. Indeed, as computers became more powerful, the number and complexity of the problems that could be addressed through numerical methods and numerical simulations increased exponentially, making problems that were inaccessible with analytical methods treatable or at least affordable.

World War II provided an important stimulus to the development of algorithms and automatic calculations. On one side of the Atlantic, we find the encryption (Germany, with the machines Enigma and Lorenz SZ40/42) and decryption machines (Poland, with the machine Bomba, and England, with the machine Colossus that

---

[3]Picone was an influential Italian mathematician, founder in 1927 of the "Istituto Nazionale per le Applicazioni del Calcolo" (National Institute for Calculus Application) the first applied mathematics institute ever created, although a computation lab had already been founded in Great Britain by E. T. Whittaker at the University of Edinburgh at the beginning of 1900. Whittaker also wrote the first complete book on numerical analysis (E. T. Whittaker and G. Robinson, *The Calculus of Observations. A Treatise on Numerical Mathematics*, London, 1924) which has remained a reference textbook on the subject for decades.

can be considered as the world's first programmable, electronic, digital computer and it was developed using the ideas of one of the most important scientists in the computer history, Alan Turing). On the other side, in United States, exceptional scientists, such as Enrico Fermi, Richard Feynman, and John von Neumann, within the Manhattan project, were struggling with calculating machines (as Brunsviga calculator and Marchant calculator) and punch-card style calculators (IBM Mark I) to help with the complicated calculations for the construction of the first atomic bomb. After the war, which highlighted the potentiality of electronic computing, there was a rapid development of digital computers mainly with the seminal contributions of Alan Turing and John von Neumann. Alan Turing, in his theoretical work *On Computable Numbers* (1936) proposed a *"universal computing machine"* that would be able to perform any imaginable mathematical computation admitting an algorithmic representation. Von Neumann, exploiting Alan Turing's idea, in his work *First Draft of a Report on the EDVAC* (1945) presented the first description of a computer architecture (known as "von Neumann Architecture" or nowadays "stored-program computer") able to execute any kind of computation accordingly to detailed instructions without the need of hardware modifications. The basic idea is that instructions and data share the same memory, and a Central Unit (the CPU in modern language) executes the instructions stored in the memory. This new type of design made obsolete the old conception of "setting-up" of a computer, that is the physical plugging and unplugging of wires on the machine (i.e., hardware modification) in order to solve different problems (that was how old computers were programmed). For selecting the appropriate set of instructions to be implemented on a computer, in 1946, von Neumann was involved as a consultant in the "stored-program" modification to the ENIAC computer, that is the first realization of his idea.[4]

Bringing together dozens of leading scientists in the Los Alamos Laboratory during World War II has led to the development of many ideas that have changed the course of science. In particular, concerning algorithms and numerical simulations, we can mention the Monte Carlo algorithm, that is a statistical method able to compute quantities of interest by the use of repeated random samples. The idea of Monte Carlo method was due to Stanislaw Ulam[5] who, in order to compute the probability to solve a complicated solitaire (very difficult to obtain only using combinatorics) had the idea of playing the game many times obtaining the probability empirically as the ratio between the number of success over the total number of games (that is an application of the law of the large numbers, see the entry Probability). This simple and powerful idea, based on try and count instead of direct calculations, obviously

---

[4]For the interested reader we suggest the classical book on the history and the development of the computer by H. H. Goldstine, *The Computer from Pascal to von Neumann* (Princeton University Press, 1972).

[5]However in his work "The beginning of the Monte Carlo method", Nicholas Metropolis (an important scientist at Los Alamos Laboratory and co-author with Ulam of the first paper about Monte Carlo method in 1949) affirms that in a conversation with Emilio Segré (Nobel Prize in physics, former student of Enrico Fermi) comes out that Fermi had independently introduced the statistical sampling technique in order to easily calculate predictions about some experimental results. Unfortunately he did not publish anything about this topic.

refined and enriched in order to be applied in more general contexts, is at the basis of an impressive series of results pertaining very different disciplines ranging from particle physics to finance. One of the first scientific applications of the Monte Carlo method is due to Ulam and von Neumann about the problem of neutron diffusion in fissionable material, obviously in the context of the development of the hydrogen atomic bomb. This idea has been implemented firstly by the FERMIAC, also named "The Monte Carlo trolley", that is an analog computer invented by Enrico Fermi to study the process of neutron transport, collision and fission using the Monte Carlo method, and later the algorithm was implemented in the ENIAC computer.

The first example of numerical computation which succeeded not only in facilitating the calculation of interesting physical quantities, but also in showing an absolutely unexpected type of behavior in a physical system came out of the forge of the Los Alamos laboratory by the work of Enrico Fermi, Stanislaw Ulam, and John Pasta with the help of Mary Tsingou in 1953. The Fermi-Pasta-Ulam-Tsingou[6] experiment is a numerical simulation of a system composed of a large number of beads connected via a linear plus a small non-linear interaction term (like springs). In the absence of the non-linear term the system's evolution is perfectly predictable and the dynamics can be decomposed in the different non-interacting vibrational modes. The scientists thought that, activating even a small non-linear interaction, the energy would flow through the various vibrational modes until reaching the equipartition of energy, i.e., after a certain time the energy of each mode would fluctuate around the same value determined by statistical mechanics. On the contrary the FPUT numerical experiment revealed a quasi-periodic behavior without equipartition of energy. The FPUT problem has a long tradition in science and only recently several ideas allowed to start understanding the reasons for the failure of energy equipartition.

Another milestone in numerical simulation is the work of B. J. Alder and T. E. Wainwright (1957) in which they obtained for the first time a phase transition, similar to the liquid-gas or liquid-solid transitions, in a hard sphere system. They simulated the evolution of 32 and 108 particles using a technique that nowadays is called Molecular Dynamics (MD)—a very important numerical simulation method that calculates the evolution of particle systems (atoms or molecules) according to Newton's laws. Unlike the Monte Carlo algorithm that is used to compute averaged quantity by means of probability distributions, MD calculates the evolution of each

---

[6]Formerly known as Fermi-Pasta-Ulam, until Thierry Dauxois in 2008 unveiled the fundamental role of Mary Tsingou in the work. Surely it was Enrico Fermi who proposed to use a computer to test a physical hypothesis about the role of a small non-linearity for the equipartition of energy (de facto inventing numerical experiments) just as surely as Mary Tsingou was the one who invented the algorithm to translate the idea of a numerical experiments in a programming language running on the computer Maniac I in 1953. There are many cases of women whose contributions to the development of computer science and technology were recognized only many years later, as the ENIAC Girls, a group of six young women who were the first programmers of the ENIAC Computer (one of them, Adele Goldstine, was involved in the "stored-program" modification of the ENIAC computer together with John von Neumann) or the female figures described in the aforementioned movie *Hidden figures* by Theodore Melfi, that allowed important progresses in the American space program. In all those cases the women did not receive any recognition for their work in their lifetimes, and only recently historians are starting to shed some light on their important role.

particle with good accuracy and exploits the idea of ergodicity (see the entry Statistical Mechanics) to evaluate average behaviors. The Monte Carlo algorithm is usually
faster than MD but it is less accurate. Indeed, in the first work on phase transition by
M. N. Rosenbluth and A. W. Rosenbluth (1954) based on the Monte Carlo algorithm
the authors reached the conclusions that *"Results obtained thus far lead us to feel
strongly that the Monte Carlo method is a useful tool for solving statistical mechanical problems, although it does not appear to be feasible to obtain detailed results
in transitions regions"*. Many years have passed since then and numerical simulation methods have become very effective thanks to increasingly clever algorithms
and increasingly powerful computers, but the distinction between statistical methods
(as the Monte Carlo algorithm) and dynamical methods (as MD) still characterizes
numerical simulations.

It is worth also mentioning the model of E. Lorenz for a minimal description of
atmospheric convection (1963) in which an unexpected and astonishing behavior
has been highlighted only thanks to numerical simulation. Lorenz was interested
to obtain a simplified model for the air mass movements in the atmosphere. He
obtained a synthetic system of three coupled non-linear differential equations whose
evolution showed a stunning behavior, i.e., two trajectories with slightly different
initial conditions after a while separated in a dramatic way (see the entry Chaos
for a further discussion). This work marked the (re)discovery (after the seminal
contribution of H. Poincaré) of deterministic chaos, and we can safely affirm that
without numerical simulations the development of chaotic dynamical systems would
not have had the breadth that it actually had.

In the past, different scientists realized the existence of some strange mathematical
objects, e.g., Weierstrass's function, a continuous function never differentiable,[7] or
the findings of J. H. Poincaré in the three bodies problem,[8] but the relevance of their
discoveries were not well understood. Merit of numerical simulations has been also
to allow for a systematic investigation able to go beyond the rigorous mathematical
studies making visible fractals (see the related entry) and chaotic behaviors, which
descend directly from the Weierstrass' function and the discoveries of Poincaré,
respectively. For instance, while fractals[9] had already been previously studied, they
attracted a massive interest of the scientific community only when it was possible
to visualize them thanks to computers. Indeed it is one thing to study the strange
properties of the Julia set (1918),[10] a different thing is, thanks to computers, to
visualize its contours and to realize that they are rather similar to the geometry
of many natural objects, from the structure of the coastlines to the shape of the
Romanesco broccoli.

---

[7] Also known as the "monster", that, in the words of J. H. Poincaré was *"an outrage against common
sense"*. The reader can see an example of such function in the entry Richardson, which was one of
the pioneers of fractal geometry.

[8] The solution of Newton equation for three bodies in gravitational interaction, which also gives rise
to chaotic trajectories.

[9] Which are tightly related to non-linear dynamics and chaos.

[10] Given the iterative map $z_{n+1} = z_n^2 + c$, the Julia set is defined as the set of the initial condition
z_0 in the complex plane such as $\lim_{n \to \infty} |z_n| < \infty$.

Nowadays, beyond the applications in physics and mathematics, numerical simulations are used in many different fields ranging from chemistry to biology, from engineering to finance and sociology. The possibility to perform experiments in which the constitutive elements of a system are decided at will is a fundamental prerogative of numerical simulations that can be hardly carried out by means of real (laboratory) experiments. Such flexibility represents a powerful mean of investigation and analysis which often allows to identify the few "key ingredients" for observed phenomena leaving aside all those features that complicate the description and can hide the important aspects, thus enormously facilitating the work of the scientists. The aforementioned Lorenz model is a perfect example, the simulation of an extremely simplified model, allowed to fully appreciate the importance of chaos in atmospheric dynamics.

It is worth here to mention the Human Brain Project, aiming at modeling and investigating, with numerical simulations, the entire human brain on different spatial and temporal scales, from the molecular level to the description of the large networks underlying higher cognitive processes. Moreover, taking advantage of the present numerical capabilities, it is possible to carry out simulations of systems that are absolutely intractable without numerical aid. For instance, nowadays it is possible to simulate at the atomic level entire biological entities, such as proteins, and with models, i.e., considering only the most important elements, one can simulate the blood flow in the arterial network or the human immune system response after an infection, with no need to explain the benefit these possibilities carry along.

The examples just discussed, as well as many other applications, show the great importance of numerical simulation as a new approach of the scientific investigation different from theoretical and experimental studies. Numerical experiments are not in contrast with classical laboratory experiments since the latter represent the final test for a theory that aims at explaining some aspects of reality, however numerical simulations have become an indispensable aid to the construction and verification of models. It is worth mentioning the application of numerical simulations to fluid mechanics, which were pioneered by von Neumann (see also the entry Turbulence). In this case the Navier-Stokes equations are considered to be a very reliable description of the fluid and numerical experiments[11] constitute an important counterpart to laboratory experiments. Indeed while the former do not allow to reach very high Reynolds number (as typically encountered in technological and natural fluid mechanics) they offer an enormous advantage with respect to the latter as they allow full access to the entire velocity field, which is not possible in the lab.

On the other hand, from a conceptual point of view, the issue of the reliability of numerical experiments in generating new knowledge has been raised, especially when numerical simulations do not have a solid counterpart in classical experiments, as, for instance, in the simulations of models for the general circulation of the atmosphere or in the simulation of the expansion of the universe, for which no real experiments can be performed. A particularly important example concerns the relia-

---

[11] In scientific jargoon called Direct Numerical Simulations, as they are a discrete but fully realiable numerical implementation of the Navier-Stokes equations.

bility of the forecasts of climate models. In this case taking (or not taking) a decision based on the results of a numerical simulation could have a huge impact on the entire human society, for better or worse. But even in less dramatic situations, for example, in numerical simulations leading to the design of a new wing of an aircraft or to the supporting structure of a bridge, the problem whether or not one can trust numerical simulations is central. When simulations are based on well-established theories (like mechanics or quantum mechanics) and the model is not too complicated, that is, it does not involve too many and dissimilar length scales and time scales, one can safely trust in the results of the simulations. More problematic is the study of social or economic problems, where sure theoretical frameworks are still lacking.

Chapter 10
Determinism

In the modern scientific terminology, the adjective deterministic is used to denote those systems whose future evolution is determined once the initial state is given. In other terms, a system is considered deterministic when in its evolution there is no room for chance, and there are no random or accidental elements that could influence system's dynamics. Although the definition just given appears to be quite clear, the concept of determinism is one of those general and fundamental ideas that, over the history, have been considered through different intellectual perspectives and disciplines each time renewing its meaning and causing debates between proponents and detractors. A brief excursus into the origins and the evolution of this concept is surely very useful in order to have a fair understanding of it and to unveil some delicate aspects. The roots of the most important philosophical ideas date back to the ancient Greeks and "determinism" is not an exception.

Just to mention a very old source, in Leucippus' fragments (fifth century B.C.) it is possible to read *"nothing happens in vain, but everything from reason and by necessity"* that can be seen as the first expression of determinism in philosophy, in agreement with the modern definition, e.g., for the Stanford Encyclopedia of Philosophy determinism is *"the idea that every event is necessitated by antecedent events and conditions together with the laws of nature"*.

In the *Timaeus* of Plato (360 B.C.), a dialog between various characters, Plato discusses the nature of the physical world and the properties of the universe and affirms that material things cannot act for a reason because their behavior is determined by external causes: there is only one way in which they can behave, given the circumstances. That is a strong affirmation of determinism: given a certain premise, the course of action of a material thing is established.

Zeno of Citium (334 B.C.–262 B.C.), the founder of stoicism, asserts that the universe is completely deterministic, nothing happened by chance but everything was determined by the fate, *"Ducunt volentem fata, nolentem trahunt"*—Fates lead those who want, they drag those who do not want to.

© Springer Nature Switzerland AG 2021
M. Cencini et al., *A Random Walk in Physics*,
https://doi.org/10.1007/978-3-030-72531-0_10

Nevertheless, other important philosophers believed that the evolution of the universe is not determined, but it is influenced by chance. For example, Lucretius in his *De rerum natura* (On the Nature of Things—first century B.C.) introduced what he called the *"clinamen"*, i.e., an unpredictable change ("swerve") in the motions of atoms, sufficient to break the causal chain of events and to restore the free will that appeared to be negated by determinism brought to its extreme consequences (i.e., considering determinism not only for material things but also for living matter).

Jumping a few centuries, we can quote Cicero's *De Divinatione* (44 B.C.), where he discusses the link between determinism and the possibility of making predictions about the future, an argument which, as we shall see later, is instrumental in the modern treatment of determinism: *"[...] if there were a man whose soul could discern the links that join each cause with every other cause, then surely he would never be mistaken in any prediction he might make. For he who knows the causes of future events necessarily knows what every future event will be"*.

Although the ideas of the Greek and Latin philosophers are fascinating, the concept of determinism received a precise and operative connotation only after the scientific revolution in the seventeeth century. We can summarize some new and revised concepts as follows:

"Causality": with Galileo Galilei this term assumes an operative meaning, *"Causa è quella, la quale posta segue l'effetto, e rimossa si rimuove l'effetto"*— Cause is such that when it is active the event occurs, when it is removed the event does not occur. Following the new causality notion, Galilei prepared his experiments investigating the laws of nature by actively exploring and manipulating natural processes;

"Geometry and Physics": with René Descartes the mathematical (and geometrical) reasoning becomes the privileged language in the description of physical phenomena: *"non alia principia in Physica quam in Geometria, vel in Mathesi abstracta"* (there is no need for other notions but mathematics and geometry in order to investigate the laws of nature). The extreme mechanistic philosophy of Descartes, in which all the natural phenomena can be traced back to mechanical behavior of the constituents of matter, cannot find a reasonable and solid explanation for phenomena as gravity or magnetism, but it has the enormous merit of considering the mathematical description of phenomena as the fundamental one;

"Laws of motion": as stated in Newton's Principia Mathematica (Mathematical Principles of Natural Philosophy) are the founding stones of the modern explanation of natural phenomena. Newton introduced the three laws of mechanics and the law of universal gravitation making the first great unification in physics: the motion of objects on Earth are governed by the laws of mechanics, and Kepler's laws can be derived by the gravitational forces and the laws of mechanics.

Of course, the scientific revolution was marked by the works of many other important scientists, such as Copernicus and Leibniz just to mention two of the most well known. What is important to highlight here is that the concepts of determinism and causality, endowed for the first time with the language of mathematics, by means of physical laws allowed to explain and also predict physical phenomena. Indeed, the mathematical description, via equations, of the dynamics of physical systems led to the possibility of making predictions on the future evolution of such systems. The fundamental second Newton's equation

$$f = ma$$

implies that once all the forces, f, acting on a body of mass m are known, we know the acceleration, a, to which that body is subjected. Then given the present state of the body (namely, its position and velocity) one can quantitatively predict the future evolution of the body by integrating the differential equation

$$\begin{cases} x'' = f/m \\ x'(0) = v_0, \\ x(0) = x_0 \end{cases}$$

where the acceleration, $a = x'' = d^2x/dt^2$, is the second derivative of the position, while $x'(0) = v_0$ and $x(0)$ are the initial velocity and position of the body,[1] respectively. After Newton we had the possibility to make quantitative predictions in a single unified framework both about the dynamics of bodies on Earth (for example, calculating the trajectory of cannon balls) and about the dynamics of celestial bodies (for example, making predictions about the position of the Moon). The success of celestial mechanics helped to strengthen trust in science as a driving force for prediction and in scientific determinism as a true possibility.

What it is usually considered as the manifesto of scientific determinism can be found in the book *A Philosophical Essay on Probabilities* (1814) of Pierre Simone de Laplace: *"We ought then to regard the present state of the universe as the effect of its anterior state and as the cause of the one which is to follow. Given for one instant an intelligence which could comprehend all the forces by which nature is animated and the respective situation of the beings who compose it—an intelligence sufficiently vast to submit these data to analysis—it would embrace in the same formula the movements of the greatest bodies of the universe and those of the lightest atom; for it, nothing would be uncertain and the future, as the past, would be present to its eyes"*. In other words, the laws of nature can allow someone (a superior intelligence, "Laplace's demon") to determine the future evolution of all the elements composing the universe once known all the forces acting on them and their initial conditions. Torrents of words have been uttered and written about this sentence of Laplace (see also the entry dedicated to him). In the following we will briefly discuss some topical questions concerning this manifesto of determinism.

As first it is important to recall that, although Laplace is usually considered as the champion of determinism, surely he was not the first who discussed determinism in a scientific or philosophic framework. Similar ideas about a powerful intelligence with a complete knowledge of forces among all elements in nature able to describe the future states of the universe with infinite precision had been proposed several times, for example, in the work of Maupertuis (1756) and Condorcet (1758). In particular

---

[1] Just for notation simplicity we consider the one-dimensional case.

D'Holbach in his book *The system of Nature* (1770) presented the idea of a "hard determinism", in which also the behavior of human beings are predetermined since they are "material things" (also their soul, *"the soul is submitted to precisely the same physical laws as the material body"*) and all the changes in material things are uniquely determined by immutable laws. Even though the idea of hard determinism for human beings appears to be quite questionable, it can be considered as a natural consequence of Laplace's idea about the possibility to predict the future state of the entire universe: *"A geometrician, who would exactly know the different forces [...] and the properties of the molecules that are moved, could demonstrate that, according to given cause, each molecule acts precisely as it ought to act and could not have acted otherwise than it does"*.

Another important consideration is that usually Laplacian determinism is scientifically justified by considering Newton's equations of motion and the fact that, under particular mathematical requirements (usually verified by these equations), once known the initial condition of the system there is a unique solution of this equation, i.e., the system evolution is perfectly and completely determined. On the other hand, Laplace's determinism must be viewed more related to philosophic traditions than about Newton's equations of motion. Actually the theorem of the existence and uniqueness of the solution of differential equations is posterior of more than 50 years to Laplacian determinism. It is therefore not wrong to consider Laplacian determinism as an heir to the classical tradition (obviously reinforced and enriched by the physics of Newton's equations) that goes from Cicero's *De Divinatione* (well known by Laplace) to the principle of sufficient reason. This last principle, stating that everything that happens must have a reason, a cause, is usually attributed to Leibniz, but the underlying idea has its origin, as usual, in the Greek philosophers (compare with the Parmenides sentences *"ex nihilo nihil fit"*, i.e., nothing comes from nothing).

Most people believe that, owing to the great predictive success achieved by celestial mechanics, Laplace has been an ineradicable proponent of pure determinism. This view is maybe too naive, considering that his famous sentence about determinism is in the opening of a chapter titled "Concerning probability" in the book *A philosophical essay on probabilities*. Why did Laplace write the most famous sentence on determinism in a book about probability theory? Laplace was a great pragmatic scientist who knew very well that to face a problem all the useful tools must be used, and although perhaps he really believed in determinism, in practice the calculation of probabilities is very useful in all those problems in which, for any reason, all the details of the system are not known, or if there are so many elements involved. Pragmatically, since we are not *"an intelligence which could comprehend all"* to proceed we must use probabilities.

The possibility to describe all phenomena through a common language is a profound human wish, dating back to presocratic philosophers. After the scientific revolution, this program found a real and concrete meaning with the construction of modern science through the language of mathematics. The increasingly important scientific advances of the nineteenth century, from thermodynamics to electromagnetism, gave new blood to determinism that became one of the principal elements leading to the birth of positivism phylosophy. In positivism any knowledge that comes from metaphysical sources is rejected and only those theories that are based on experimental evidence are accepted. Supported by a solid trust in logical reasoning and guided by the experimental method, the positivists thought they could give a mathematical description to all phenomena, whether natural, biological, or social. One of the founder of the positivism, August Comte, affirms that just as the physical world is governed by laws so it is with society.

An enlightening example of the impact of determinism and positivism in social sciences is given by the positivistic criminology for which causes of criminal behavior must be sought among sociological, psychological, and overall biological causes. The champion of such an approach has been Cesare Lombroso who, speculated about abnormal behavior, established biological specificities for criminals. Lombroso thought that some criminal characteristics were hereditary and identifiable through a scientific re-edition of physiognomy. Nowadays, it is well evident that some of the positivistic social theories are too simplistic and naive,[2] however the nineteenth-century positivism had the merit of considering means of investigation similar to those of the natural sciences also in social sciences. This is certainly a positive aspect, less positive is to use a very precise scientific terminology in a context in which these margins of precision do not exist.

Concerning physics, positivism led nineteenth century scientists to believe that through mathematical language the human intellect could have access to all natural phenomena: the reaching of a complete knowledge of the laws of nature was no longer in question, it was only a matter of time.[3] In this perspective also reductionism played an important role. Reductionism can be viewed both as an operative methodology, i.e., the study of a system considering only the simplest case, i.e., reality can be understood once the fundamental laws governing the behavior of the ultimate constituents of the matter are discovered. Of course, such an attitude has an undoubted operational value (starting from Galileo Galilei) that has led to the progress of science as we know it today, on the other hand the concept of reductionism, has shown more and more limits as some crucial discoveries of the twentieth century have come to light, see the entry *Laws, levels of description and models* for a brief discussion.

From the point of view of modern physics, the practical and conceptual relevance of determinism does not appear particularly significant. In fact, in the twentieth

---

[2] Although, even recently, questionable studies are still being proposed, focusing on the relationship between physical characteristics and personality traits, e.g., in the 1990s some authors affirmed to be able to detect male homosexuality by looking at the pattern of whorls in the scalp, or more recently a machine-learning study affirms that an algorithm could detect sexual orientation "more accurately than humans" (in 81% of the tested cases for men and 71% for women).

[3] Actually, some years after, the great physicist Paul Dirac wrote: *"The underlying physical laws necessary for the mathematical theory of a large part of physics and the whole of chemistry are thus completely known, and the difficulty is only that the exact application of these laws leads to equations much too complicated to be soluble"*. Actually very often *complicated* means *impossible to treat*, and therefore having a knowledge of the correct equations of a given phenomenon is not conclusive, and it is mandatory to introduce approximate methods so to obtain an explanation of the main features of complicated systems, see also the entry Laws, levels of description and models.

century at least two fundamental discoveries undermined determinism as presented by Laplace: quantum mechanics and deterministic chaos.

At the beginning of the last century, the need for a description of some phenomena at microscopic level, such as the black-body spectrum, i.e., the radiation emitted by a physical body that absorbs all incident electromagnetic radiation, or the photoelectric effect, i.e., the emission of electrons from materials when irradiated with ultraviolet light, led to the development of quantum mechanics which gave a first shock to the concept of determinism. Actually the uncertainty principle of quantum mechanics affirms that there is a limit to the precision with which position and speed of a particle can be simultaneously measured: if one measures the position of a particle with good accuracy, then he will have very limited knowledge of its velocity, and vice-versa. Moreover, Newton's laws of motion failed to describe particle dynamics in the microscopic world were replaced by the Schöedinger equation describing the evolution of the so-called wave function from which one can only determine the probability of finding a particle in a certain position at a given time, leading chance to enter physical laws at a fundamental level.

Somebody might still believe that, although there are some problems in the microscopic world, if classical mechanics holds there is no reason to doubt about the central role of determinism. Actually the discovery of deterministic chaos frustrated also this possibility (see the entry Chaos). Various works from the middle of the last century showed that even in systems governed by simple differential equations long-term predictions are actually impossible. This happens when coping with non-linear systems in which small uncertainties on the initial state of the system are exponentially amplified as time proceeds, so that the motion of two initially very close states of a system quickly become completely independent (sensitivity to initial conditions). In these cases the system is said to be chaotic. Therefore, if the initial condition of a chaotic system is not known with infinite precision, it no longer makes sense to speak of long-term prediction. Strikingly, this apparent oddity belongs also to the discipline that was considered the triumph of determinism, namely, celestial mechanics, as it was actually discovered by Poincaré (see the dedicated entry).

The idea that a perfectly deterministic system, if chaotic, cannot be predictable could generate some confusion. So does determinism not imply predictability? Even famous philosophers tend to confuse the two terms: *"Scientific determinism is the doctrine that the state of any closed physical system at any future instant can be predicted"* (Popper 1992). To shed some light on this thorny topic we need to point out that the two terms belong to two different levels of description. The term *deterministic* concerns the nature of a given system, it is an ontic (linked to the essence) property of the system, while the term *prediction* concerns the operative ability to forecast the future evolution of a system, predictability is an epistemic (linked to the knowledge)[4] characteristic of a system. It is possible to have deterministic systems that are not

---

[4]We stress that the impossibility to predict is indeed linked to our limitation to know with infinite precision the state and/or the parameters entering the laws describing a given system, again the problem is that we are not *"an intelligence which...."*, back to Laplace. Hence, as Laplace had very clear in mind, the necessity of probability also in a deterministic context.

predictable, at least for long times as in chaotic systems, and non-deterministic systems that are predictable, for instance, in systems with many components and stochastic noise, where one can use the results of the limit theorems of probability.

Chapter 11
Einstein

Albert Einstein (1879–1955) is recognized as one of the most important scientists and thinkers in the history of mankind. His ideas radically changed our view of space and time. His portraits, personality, and even mathematical formulas (such as $E = mc^2$) entered the collective imagination, becoming as popular as Beatles's songs or Wharol's artworks. He received the Nobel Prize in Physics of 1921 (assigned in 1922) for *"his services to theoretical physics, and especially for his discovery of the law of the photoelectric effect"*, and he was chosen as the Person of the Century from the magazine Time. A huge amount of literature exists, on both Einstein's life and scientific results. In the following, after a succinct review of his biography, we focus on a few salient aspects of the beginning of Einstein's scientific career (until 1905), which is rooted in a strong interest for statistical mechanics, a fact frequently ignored in popularization essays.

Einstein, born in Ulm (Germany) in a Jew family, received his higher education in Zurich (Switzerland) at the Swiss Federal Polytechnic School (now ETH) between 1896 and 1900. For several reasons he did not find the ETH environment congenial, he cut classes and worked a lot on his own. As a consequence, once graduated he found himself without any assistantship or recommendation: for some years he lived as a young graduated student without a stable work, sustaining himself by private tutoring and temporary teaching positions. His life at that time was not very different from that of the many young graduated students around the world today. Difficulties also arose because his relationship with Mileva Marić, a fellow student at the ETH, was not approved by his parents. His lack of a position prevented their marriage until 1903, 1 year after he had found a stable employment at the Swiss Patent Office. It is important to recall that during those years Einstein worked to his Ph.D. thesis, whose subject was initially thermoelectricity and then changed into molecular kinetics.

Interestingly, Einstein's status of precarious science worker ended only in 1909, when he was appointed associate professor at the University of Zurich, under the recommendation of Kleiner, his final Ph.D. advisor. In the next years he became full professor in Prague and then in ETH Zurich, where he taught analytical mechan-

© Springer Nature Switzerland AG 2021
M. Cencini et al., *A Random Walk in Physics*,
https://doi.org/10.1007/978-3-030-72531-0_11

ics and thermodynamics. After a few years he moved to Germany where he rapidly reached top positions such as director of the new Kaiser Wilhelm Institute for Physics and president of the German Physical Society. In 1919, after the striking experimental confirmation of the theory of general relativity by Sir Arthur Eddington, Einstein's celebrity reached worldwide and popular acclaim, with articles appearing on international media such as the leading British newspaper The Times (with the banner headline "Revolution in Science—New Theory of the Universe—Newtonian Ideas Overthrown"). Furthermore, as previously mentioned, 3 years later he received the 1921 Nobel Prize in Physics.

After years of further scientific achievements and important travels around the world (mainly in US and in Far East, Japan included), he was obliged to leave Germany as a refugee. In 1933, in fact, the new German government led by Adolf Hitler passed laws barring Jews from holding any official positions, including teaching at university. Thousands of Jewish scientists were suddenly forced to give up their jobs, apparently without any important protest raised by their colleagues. The Nazi propaganda targeted Jews intellectuals and their books, together with many other "un-German" books, were burned. Einstein, in a letter, wrote *"I must confess that the degree of their brutality and cowardice came as something of a surprise"*. Einstein first moved to Belgium and, at the end of 1933, took up a position in Princeton, at the Institute for Advanced Study. In 1935, he finally applied for US citizenship and, scientific travels apart, he stayed there until his death in 1955. At the Institute, Einstein developed friendship and collaboration with many scientists including John von Neumann (see the related entry in this book) and Kurt Gödel. At the beginning of World War II, Einstein come to know from Szilárd and Wigner of the concrete possibility of ongoing research for an atomic bomb in Germany. A famous letter followed, signed by Einstein and Szilárd, addressed to US President Roosevelt recommending the US to seriously consider nuclear weapons research. Einstein, however, was a pacifist and later in his life wrote to Pauling *"I made one great mistake in my life— when I signed the letter to President Roosevelt recommending that atom bombs be made; but there was some justification—the danger that the Germans would make them"*. Just before his death, together with Bertand Russell and other important scientists, he authored the famous Russell-Einstein Manifesto issued on 9 July 1955 pursuing the following resolution *"In view of the fact that in any future world war nuclear weapons will certainly be employed, and that such weapons threaten the continued existence of mankind, we urge the governments of the world to realize, and to acknowledge publicly, that their purpose cannot be furthered by a world war, and we urge them, consequently, to find peaceful means for the settlement of all matters of dispute between them"*. The manifesto gave birth, 2 years later, to the Pugwash Conferences on Science and World Affairs, an organization that in 1995 was the recipient of the Nobel Peace Prize *"for their efforts to diminish the part played by nuclear arms in international politics and, in the longer run, to eliminate such arms"*.

In his lifelong production of over 300 scientific papers Einstein investigated several different subjects, including thermodynamics, statistical mechanics (classical and quantum), special and general relativity, cosmology, and much more. Notwithstanding the abundance of his whole scientific life, it is usually accepted that at (almost) the beginning of his career, in a single year, Einstein produced an exceptional concentration of novel ideas. Indeed in 1905 Albert Einstein published four fundamental papers. For the value of these four papers 1905 is often referred to as "Einstein's annus mirabilis", that is "marvelous year". Two of these papers represent the foundations of the successive great achievements of Einstein in revolutionizing basic concepts such as space and time: one about special relativity and one about the energy-mass equivalence (the latter is, ideally, the follow-up of the preceding one). A third paper, the one on photoelectric effect, represents a fundamental contribution of Einstein to the nascent quantum theory, and also constitutes the main motivation of his Nobel prize. Much less known to the great public is the fourth paper, the one about Brownian Motion (see the corresponding entry). This—from the point of view of non-specialists—may seem a minor work. On the contrary, it is a pillar of statistical mechanics (see the corresponding entry) which is the science of understanding thermodynamic laws—or macroscopic laws in general—merging together the laws of mechanics and probabilistic concepts and represents one of the main branches of theoretical physics (at that time as well as nowadays). This work also contains the seeds of future developments of the theory of stochastic processes, in probability, which has widespread applications, for instance, in chemistry, economy, finance, etc.

To the largest public, Einstein is known as the scientist who deeply changed our perception of time, connecting it to space in an intrinsic way. The fact that time is inextricably related to space is stated in Einstein's 1905 works on special relativity: in that works Einstein needed to guarantee true (as apparent from experiments) the principle of constant speed of light in the vacuum, and for this reason time must adjust to the speed of the laboratory where the measurement is performed. Later, in 1915, Einstein generalized special relativity to systems with acceleration, creating general relativity, a theory where space-time is not flat but can be curved according to the properties of matter inside it. Space-time curvature explains gravity. Einstein's field equations for general relativity are rich and complicate, albeit apparently simple and compact. They contain many predictions, including the existence of black holes and the deflection of starlight by the Sun. The latter phenomenon was experimentally confirmed by Sir Eddington during the solar eclipse of May 29, 1919. General relativity concepts such as black holes or the curvature of space-time (and later extrapolations such as wormholes and time-travels) are today extremely popular and appear in best-seller fiction books and even in Hollywood movies with very large audiences. Sometimes these concepts are used with extreme accuracy and effectiveness, as in the recent Interstellar blockbuster directed by Christopher Nolan (2014).

Apart from relativity, Einstein is also well known for his explanation of the photoelectric effect, another 1905 paper, where he put forward the idea that light is not a continuous wave but rather it travels in discrete (quantized) packets. The formalization of this idea clarified previous work by Planck on the black-body radiation, that is the light emitted by a body only because of its internal molecular agitation due to its temperature, and not due to any kind of reflection. Einstein's theory was confirmed experimentally by Millikan a few years later. The contribution to the understanding of the photoelectric effect constitutes the motivation of his 1921 Nobel prize and is one of the founding pieces of quantum theory. Wave-particle duality is another key revolutionary concept which made through the boundaries of the community of specialized scientists and now belongs to the imaginary of the great audience. About this, it is worth mentioning the celebrated scientific and philosophical debate that involved Einstein and Bohr in the late 20s and 30s, concerning the probabilistic interpretation of quantum mechanics.

The non-specialist public is usually unaware that the first scientific interest of Einstein—before relativity and the light quantization—was understanding the microscopic origin of "heat". Apparently the idea of understanding the mechanical foundations of thermodynamic laws—that is statistical mechanics—may seem not as spectacular as changing our concepts of light or space-time. Certainly, it does not entails ideas which can be easily exploited for fiction or science popularization. However, Einstein's interest for statistical mechanics was not secondary. On the contrary, most of his scientific works in the beginning of his career (three of his first five papers between 1901 and 1904) were devoted to it. Most importantly, as anticipated, one of the four works of the "annus mirabilis" (1905) was on Brownian Motion, that is the motion of a grain of pollen continuously kicked by surrounding invisible water molecules. In 1905, Einstein also wrote his inaugural dissertation for the Ph.D., strictly related to the Brownian Motion paper (the content of this dissertation was published as an article only in 1906). Speaking about it, Einstein's friend and biographer Abraham Pais notes that it *"has more widespread practical applications than any other paper Einstein ever wrote... the thesis, dealing with bulk rheological properties of particle suspensions, contains results which have an extraordinarily wide range of applications. They are relevant to the construction industry (the motion of sand particles in cement mixes), to the dairy industry (the motion of casein micelles in cow's milk) and to ecology (the motion of aerosol particles in clouds). Einstein might have enjoyed hearing this, since he was quite fond of applying physics to practical situations"* (see the entry Mesoscale systems).

To understand the great interest of Einstein for statistical mechanics it is useful to recall a sentence at the beginning of the second 1902 paper: *"Great as the achievements of the kinetic theory of heat have been in the domain of gas theory, the science of mechanics has not yet been able to produce an adequate foundation for the general theory of heat, for one has not yet succeeded in deriving the laws of thermal equilibrium and the second law of thermodynamics using only the equations of mechanics and the probability calculus, though Maxwell's and Boltzmann's theories came close to this goal. The purpose of the following considerations is to close this gap"*. Among specialists, e.g., physicists, the first Einstein's scientific interests and works are well known and recently they have been discussed in a very detailed and interesting way. Studying these works it is even possible and fascinating to connect Einstein's interests for statistical mechanics to his later more popular works on light quantization and special relativity.

The first two papers written by Einstein investigated a molecular force law. The successive three works concern the foundation of kinetic theory and thermodynamics. They constitute a foundation for the 1905 work on Brownian Motion. It is also explicit the connection of these first interests with the 1905 paper on quanta, particularly in the third paper on kinetic theory, where the black-body radiation was explicitly discussed. The relation of these papers with the much more famous investigation on relativity is certainly more subtle. However, at the beginning of the twentieth century, electromagnetism and thermodynamics posed to physicists a similar question, that is how they could be reconciled with the long-dominant mechanical view of the world. For instance, a mechanistic foundation of electrodynamics (through the notion of a mechanical aether, the medium over which electromagnetic waves propagated) was being challenged by a new perspective with an electromagnetic basis for mass that would itself underlie mechanics. In parallel, Planck and Ostwald defended thermodynamics as a branch of physics that could exist without grounding in mechanics. The atomic hypothesis itself, closely related to a mechanistic description of nature, did not enjoy universal acceptance. Einstein, on the contrary, was quite impressed by the ability of mechanics to explain a wide range of phenomena, by *"the achievements of mechanics in areas which apparently had nothing to do with mechanics: the mechanical theory of light and above all the kinetic theory of gases"*. Even more interestingly, in the search for the discovery of a law reconciling mechanics and electrodynamics, he came *"to the conviction that only the discovery of a universal formal principle could lead us to assured results. The example I saw before me was thermodynamics"*. These words seem to allude to a conceptual analogy between the principles of thermodynamics and those of relativity.

Thermodynamics, at that time, was the ground of an important debate involving the possibility to connect two different levels of reality. Great German scientists of that time expressed stronger or weaker doubts about the idea of a mechanistic/atomistic foundation for thermodynamics: the most important were Ostwald, Mach, Hertz, Kirchoff, and Planck, divided among those convinced that the task was wrong, pointless, or simply insurmountable, even if interesting or correct in principle (see the entries Atoms and Brownian Motion for a discussion). One approach (sustained, e.g., by Ostwald and opposed by both Boltzmann and Planck) was to replace mechanistic models made of masses and forces with models only involving transformations of energy ("energetics"). Also outside the German world the situation was not much different: for instance, the reading of Poincaré's *Science and Hypothesis*, which Einstein found very interesting, contributed to this general skepticism toward mechanistic foundations of other fields, such as thermodynamics or electrodynamics. The chief supporter of the opposite view was, of course, Ludwig Boltzmann (see the related entry). In a letter to Mileva Marić of September 1900 Einstein wrote *"Boltzmann is splendid... I am firmly convinced of the correctness of the principles of the theory"*. Einstein also sent a copy of his first molecular force paper to Boltzmann. Boltzmann believed in mechanics as the foundation of physics and certainly made use of atomic models in most of his works. In later Autobiographical Notes Einstein wrote—referring to Boltzmann's *Gas Theory*—that, as a student, he was greatly impressed by the kinetic theory of gases and the connection it offered between viscosity, heat-conduction, and diffusion of gases *"which also furnished the absolute size of the atom"*.

Even if less spectacular from the point of view of popularization, Einstein's theory of Brownian Motion has had an impressive conceptual value and it is somehow as revolutionary as his more well-known 1905s investigations on quanta and special relativity. This theory gave experimentalists sudden access to the world of atoms, which up to that time was impossible to reveal. The problem is easily explained: atoms are too small to be seen by means of an optical microscope. Statistical mechanics, before Einstein, gave many suggestions on how to connect microscopic quantities related to size (or density) and speed of atoms to macroscopically observable quantities, such as pressure or temperature. Such methods of connection, however, were based again on the measurements of fluctuations (such as fluctuations of energy) which—in a macroscopic body—are usually always too small to be detected. Einstein found a kind of "fluctuation" which can be observed: in fact—in principle—it can be very large. This fluctuation is the distance traveled by a suspended particle, small but yet visible at the microscope, such as a pollen grain on the surface of water. This particle, even if huge with respect to the atoms, moves because of the effect of the small (huge in number) incessant kicks of the surrounding atoms or molecules. The result is a random motion whose absolute distance from the starting position grows in time in a very peculiar way, very different from the law governing a particle that goes straight with constant velocity. Einstein predicted the correct kind of motion, that is known as "diffusion", and also the connection between its "speed" (which is actually the diffusivity constant) and quantities related to the microscopic world such as the Avogadro number. In a sense (with a lot of important differences, of course) this is similar to understanding the number (or size) of players in a football match by looking only at the movement of the ball for a very long time. A subsequent experiment by Perrin in 1909 confirmed Einstein's predictions: this is usually considered the first experimental evidence of the existence of atoms.

Chapter 12
Entropy

Entropy is a measure of disorder, this is the popular version. But what does disorder really mean? And why is it useful in physics? A famous anecdote tells that in 1939 Claude Shannon (1916–2001), the father of Information Theory, as well as of many fundamental concepts in the modern theory of computers, visited John von Neumann (see related entry), one of the greatest scientists of the twentieth century, asking for suggestions about a good name for the new information quantity he was inventing (see the entry on Information Theory). John von Neumann replied that a good name was *"entropy"*, first because it has basically the same mathematical definition, second because *"nobody knows what entropy really is, so in a debate you will always have the advantage"*. In fact, even among physicists, there is a widespread sense of obscurity coming from this word.

The ambiguity surrounding entropy originates from a series of real problems. First, it is a quantity which is not easily measured. Physicists usually grasp the meaning of a physical quantity by learning how to measure it: they look for an *operational definition*. How much is the entropy of the water in this glass? Is there any instrument that can measure entropy? Not exactly. Ok, is there any procedure or recipe to measure entropy? Yes but it is a complicate one. If one has not access to absolute zero temperature (a condition in principle unattainable and even very difficult to approach), then only *changes* of entropy can be measured, from one state of water to another state, for instance, when increasing or decreasing its temperature. Imagine that you know the entropy of a glass of water at a given temperature, if you want to know its entropy at a different temperature you need to perform certain measurements during the transformation from the initial to the final temperature. The change of entropy during this transformation is measured by means of a calorimeter, a thermometer, and the use of Clausius definition, which is discussed below. Even with these instruments, however, entropy changes are properly measured only by means of very slow transformations: the slower is the transformation, the more precise is the measurement. The reader understands that this recipe is not very convenient and does not help to grasp the concept entropy.

© Springer Nature Switzerland AG 2021
M. Cencini et al., *A Random Walk in Physics*,
https://doi.org/10.1007/978-3-030-72531-0_12

From the difficulty of the operational definition, a second important problem descends: the entropy of a substance is hard to be related to other physical properties of that substance. This point is important for theory: relations (or their formal counterparts, i.e., equations) are essential to make predictions or explanations of phenomena. In most of the cases, physical quantities can be *reduced* to elementary quantities that characterize the fundamental units of a substance or a body. For instance, this is the case of energy. In a theory for a particular model system, energy has a well defined formula in terms of all the essential variables of this system, such as positions, masses, and velocities of all the atoms. Unfortunately, this is not always true for entropy. At the level of large macroscopic bodies—which are not undergoing too fast transformations and therefore are well described by a few (slowly changing) "thermodynamic" variables—entropy can be put in formal relation with other macroscopic properties of the body such as temperature, pressure, density, etc. But if one considers small volumes and fast transformations, such a formal apparatus is not useful. In other terms we can define the energy of an atom but not its entropy. Entropy therefore is a well formalized concept only for some physical phenomena or at some "level of description" (see the entry Mesoscale systems). Even more deep and tormented is the relation between entropy and statistics, which is the only framework where one can appreciate the idea of entropy as a measure of disorder.

Possibly, the best way to understand the concept of entropy is through the history of its discovery, in the nineteenth century. One the most fascinating aspect of entropy is that the scientists that discovered it did not immediately caught its interpretation in terms of disorder. The name itself is not directly related to disorder. Entropy means "transformation content" from Greek particles "en" (="in") and "trope" (="turning" or "transformation"). Both the concept and the word were introduced by the German physicist Rudolph Clausius (1822–1888) in a series of works between 1850 and 1865. Clausius was interested in understanding why a heat engine is not able to convert all heat into work, but some (often large) part is lost and represents a significant wastefulness of resources: a very practical problem with a deep theoretical origin.

The investigation of the dynamics of heat at that time was a young science. In fact "thermodynamics" was born less than 30 years before, thanks to the french military scientist Sadi Carnot (1796–1832) who published the first thermodynamics book in 1824. Since the book was too abstract for engineers and too approximate and informal for physicists, it was initially ignored by most of the scientific community and rediscovered more than a decade later. In his book Carnot wanted to understand how to improve existing heat engines, at that time mainly steam machines. In doing so, Carnot reached the outstanding result of translating a very applicative problem into a conceptual one, creating an ideal model and catching the essential ingredients to investigate it: two reservoirs at two different temperatures and super-simplified transformations among them. One of the most important principles stated by Carnot was that some caloric (basically the substance representing heat in the early days of thermodynamics) must be lost in the engine, or equivalently that all the caloric entering the engine cannot be entirely transformed into useful work. Reducing this waste of caloric could improve the efficiency of engines, but even in the ideal machine

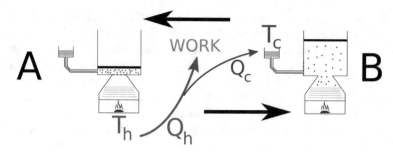

Fig. 12.1 Sketch of Carnot ideal engine, reduced to only two transformations for the sake of simplicity. In fact a correct understanding of the Carnot engine requires a more complicate treatment, but here we are keeping it to its salient ingredients

designed by Carnot there was an irreducible minimum loss, and therefore a limit to the maximum efficiency.

Between the early ideas of the French Carnot and the enunciation of general thermodynamics principle by the German Clausius, an English physicist intervened, James Prescott Joule (1818–1889), who—with a series of fundamental and undoubted experiments—demonstrated the equivalence between heat and mechanical work. Such an equivalence, totally missing in Carnot theories (for him heat was a substance, the caloric mentioned before, *falling* from hot to cold bodies), seemed to contradict even more the empirical evidence that a large part of heat is not transformed into work but inexplicably lost somewhere.

In its papers Clausius restated the principle of maximum efficiency enunciated by Carnot in more general terms, introducing a new physical quantity—called entropy only in 1865—which is equal to the ratio between the heat Q and the temperature T: the relevance of this quantity Q/T is in the fact that it must be the same if measured along all possible (slow) transformations connecting two states of a substance, where a "state" can be identified by few properties of the substance itself, for instance, (in simple gases) temperature, density, or pressure. Thanks to this concept Clausius could formalize the Carnot engine model and immediately derive its efficiency. With a little effort the reader can understand this reasoning and grasp one of the most important and elusive concepts of physics. The Carnot engine consists of repeated identical cycles: each cycle is made of a first transformation from state A to state B and then a second transformation going back from B to A, along a *different* path, see Fig. 12.1.[1]

In the transformation $A \rightarrow B$ an amount of heat Q_h goes from a source of energy at high temperature T_h into the machine. In the back transformation $B \rightarrow A$ a different amount of heat Q_c is released (i.e., dissipated) from the machine to a reservoir at low temperature T_c. Even if the two transformations ($A \rightarrow B$ and $B \rightarrow A$) are

---

[1]Basically all engines produce energy (e.g., movement) by means of repeating a given cycle of operations: for instance, the piston in most of the internal combustion engines (such as in cars) repeats indefinitely a cycle of four successive phases.

different, they connect the same states and therefore Clausius principle tells that the entropy changes must be equal: $Q_h/T_h = Q_c/T_c$. Remember now that the machine produces work thanks to the fact that heat absorbed is larger than heat released: the produced work is, indeed, the difference $Q_h - Q_c$ between energy entering (for instance, in the fire creating the steam) and energy lost (heat going in the surrounding air). Then a larger amount of work—or a better efficiency—could be achieved by reducing Q_c. Unfortunately, once T_h, T_c and Q_h are fixed, Q_c cannot be reduced, because the above Clausius principle or entropy equivalence nails it: the identity $Q_h/T_h = Q_c/T_c$ is equivalent to say that $Q_c = Q_h T_c/T_h$: there is no way to change Q_c independently of the other three quantities. The entropy equivalence also allows us to immediately compute the efficiency of the engine, based only upon the knowledge of the temperatures T_h and T_c, confirming the Carnot intuitions.[2] A few more very simple calculations (it is a nice exercise for the reader) show that, remarkably, worst efficiencies correspond to cases where entropy changes are *larger* than Q/T. If the trasformations are not slow enough, careful enough, delicate enough, they will create more entropy, which implies wasting more heat. In fact Clausius also introduced the concept of entropy production, that is the involuntary creation of entropy preventing real (not infinitely slow) engines from achieving the ideal efficiency of the Carnot model. The fact that entropy can be unintentionally created but not involuntarily destroyed is an empirical fact and is another way to formulate the second principle of thermodynamics, expressed by Clausius as the impossibility of spontaneous heat flow from cold to hot reservoirs and—in the same years—by Lord Kelvin (William Thomson, first Baron Kelvin 1824–1907) as the impossibility to entirely convert heat into work (for a more detailed discussion of this principle see the entry Irreversibility). Therefore, the concept of entropy in thermodynamics has two intertwined uses: on one side it provides a simple and precise rule for heat exchanges (entropy change is the same in all slow transformations connecting the same two states), on the other side it helps to formalize the limits of thermodynamics in a single inequality (the entropy change in a generic transformation cannot be smaller than that realized in a slow one connecting the same states).

The contributions of Clausius to thermodynamics include a first hint about the future interpretations of entropy in statistical mechanics (see the entry Boltzmann). In fact he interpreted the irreducible waste of heat (Q_c in the example above) as heat converted—during the transformation $A \rightarrow B$ into "interior work", i.e., work *"which the atoms of the body exert upon each other"*, a work (or energy) which—during the reverse transformation $B \rightarrow A$—can only be released outside and is not usable as real work. At that time the idea of matter made of atoms was accepted only by a part of the scientific community (see the entry Atoms), certainly including Clausius, who in previous years had interpreted other physical phenomena in terms of atoms. One of the most vigorous advocates of the existence of atoms in that period

---

[2]Indeed if we call $W = Q_h - Q_c$ the work produced, the maximum efficiency is simply the ratio between the work produced and the heat provided to the engine W/Q_h which, thanks to the identity $Q_h/T_h = Q_c/T_c$, can be written as $W/Q_h = 1 - Q_h/Q_c = 1 - T_c/T_h$.

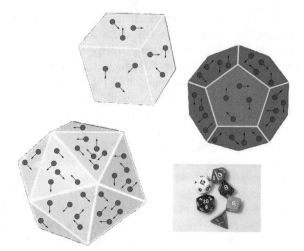

Fig. 12.2 For Boltzmann a gas is like a die with many faces, such as those used in role-playing games. The number of faces changes with its thermodynamic state, for instance larger density corresponds to more faces. A larger number of faces corresponds to a larger disorder, or entropy

was the Austrian physicist Ludwig Boltzmann, who in fact was the first to realize the fundamental connection between thermodynamic entropy and atomic disorder.

In his approach to the foundation of statistical physics, Boltzmann contributed to understand—in terms of atomic properties—both faces of entropy: how it quantifies internal disorder of matter and why it cannot spontaneously decrease. For both aspects of the problem Boltzmann made use of statistics and its mathematical counterpart, probability. Probability (see related entry) had been formalized in mathematics only in the previous 100 years, and was usually not applied to physics, with few exceptions, in particular Maxwell's contribution to the kinetic theory of gases (see the entry Maxwell), strictly related to Boltzmann's work (see also the entry Laplace) (Fig. 12.2).

Starting from the assumption that a macroscopic body is made of an enormous number of atoms or molecules, Boltzmann realized that it makes no sense to study or describe the behavior of each individual molecule. This would be like trying to understand some aspects of the population of one country by describing the behavior of each inhabitant. On the contrary, it is much more useful to consider broad categories and analyze what percentage of the total population belongs to a certain category: for instance, it is not interesting to make a list of the height of each person, while it is much more meaningful and useful saying that "10% are taller than 1.80 m, 20% are between 1.70 and 1.80 m etc.". Measuring the height of a randomly picked person would be similar to throwing a loaded die where each face has a certain probability and corresponds to a certain height category. This is—basically—the concept of statistical distribution and Boltzmann made it a powerful tool in physics. Boltzmann understood that a gas in a certain thermodynamic state, that is at given temperature, volume, and pressure, does correspond to several similar configurations of molecules. Those configurations of molecules can be categorized and associated with probabilities, like the sides of a die. Basically, according to Boltzmann's view, each thermodynamic state of gas is similar to a die with a very large number of sides, each corresponding to a particular configuration of molecules (as in Fig. 12.1). Molecules move and collide with the walls or among themselves, rapidly and incessantly changing their configuration: this is like throwing and throwing the die. Building upon this powerful idea, through examples and calculations, Boltzmann understood that thermodynamic entropy increases when the number of sides of the die increases. In fact a gas is in a state of larger entropy when that state can be realized with a larger number of possible configurations of its constituents (microstates). The number of possible microstates, say W, is basically a measure of how disordered the gas is. Technically, Boltzmann formula for entropy (stated in words by Boltzmann between 1872 and 1875 and formally written by Planck in 1900) says that the entropy S is proportional to the *logarithm* of the number W of microstates, that is the number of sides of the die,

$$S = k_B \log W,$$

being k_B a fundamental physical constant (later called "Boltzmann constant" in his honor).[3] Boltzmann formula above (which has become his epitaph) is fundamental not only as a conceptual bridge between the microscopic and the macroscopic worlds (invisible atoms and substances under our naked eyes). It is an exceptional computational tool to get predictions: the knowledge of the mechanical properties of certain kind of molecules can be used to understand how many sides the die has and, therefore, how its thermodynamic entropy behaves, i.e., how heat is adsorbed or released by a real substance. This leads to immediate applications for engineers and chemists and stands at the basis of all studies in statistical mechanics of the twentieth century.

In his investigations, Boltzmann also found a way to understand the fact that entropy cannot spontaneously decrease. He achieved such a result by the introduction of an equation that governs the time evolution of the probability of certain properties of the molecules in a gas (positions and velocities). Boltzmann exploited again his idea of entropy as the number of faces of the die (with important differences, noticeably he also included in the description the possible states of a single molecule), and calculated how entropy changes in the gas described by his equation. In this procedure he obtained via his mathematical apparatus exactly the principle

---

[3]We recall that the logarithm (to a certain base) of a given number gives the exponent of the base necessary to get that number. For instance, log 100 to base 10 is 2, because $10^2 = 100$. Notice that the logarithm of a number gives a result which increases if the number increases. So, log W it is a different way of counting the number of microstates. Note that the log operation is essential to get the following result: when two systems, with W_1 and W_2 microstates, are merged together, a new system is created with a number $W_1 W_2$ of possible microstates; the logarithm of this number is $\log(W_1 W_2) = \log(W_1) + \log(W_2)$, i.e., entropy of the merged system is exactly the sum of the entropies of the two separated systems. Of course if interactions between the two systems are important, things are more complex than this.

postulated by Carnot, Clausius, and Kelvin, i.e., that entropy cannot spontaneously decrease. Understanding this point in simple words requires a little more effort: we dedicate to it a part of the entry Irreversibility.

As mentioned before, the concept of entropy gained new life with the birth of information theory (see the entry Information Theory), in the 40s of the twentieth century, thanks to the work of Claude Shannon. Shannon introduced a whole new framework to study and explain what happens when people or machines communicate among each other. He wanted to avoid entering complicate issues such as "meaning" or "language". In fact his main aim was to rationalize and formalize what happens in communication channels, a concept which includes telephone or telegraph lines, as well as spies sending strategic information from behind enemy lines. Thanks to its high level of abstraction, Shannon's theory is considered as one of the main building blocks of today computer science. In order to appreciate how fundamental is his contribution to our everyday life, it is sufficient to recall that—in its seminal 1948 paper *A Mathematical Theory of Communication*—Shannon invented the word "bit". Entropy appears in Shannon's work as a measure of information content in long sequences of characters, such as books or telegraphic transmissions. To be more precise Shannon considered the entropy of *sources* of information, i.e., systems that generate the sequences. Entropy as information content was defined, by Shannon, with a formula which is substantially equivalent to Boltzmann's formula for entropy in statistical mechanics. Instead of configurations of a gas, Shannon considered possible strings produced by the communication source. One of the main aims of Shannon was to distinguish between strings with a lot of redundancy, i.e., with a lot of wasted characters, and sequences where each character is important and cannot be lost. The sequence "$aaaaaaaaaaaaaaaaaaaa$" is very redundant since it can be replaced by something else, for instance, "$20 \times a$" which is a shorter message. On the contrary the sequence "$djoijqweknncklskdjdkajasdkjhbab$" is hardly replaced by a shorter equivalent string. The immediate application of Shannon's entropy is, as already seen in the examples above, a measure of how a sequence can be compressed: compression is useful to save time or resources in the transmission (see also the entry Complexity Measures). Strings generated by a source with a high entropy are more difficult to be compressed than strings coming from a low entropy source. In fact an estimate of entropy can be easily obtained by means of a standard (lossless) compression software in our pc's. The deep relation between the entropy in statistical mechanics and that in information theory is due to the fact that both entropies are (logarithmic) counters of the number of microstates (for Boltzmann) or sequences (for Shannon).

This idea of relating entropy to information was prefigured by Maxwell in some letters written between 1867 and 1871 and then presented in his book on thermodynamics titled *Theory of heat* (1872). Maxwell conceived a thought experiment to understand the second principle of thermodynamics at the level of atoms in a gas. He imagined a gas in a container divided into two compartments by a middle wall. A very small supernatural being (called demon by Lord Kelvin a few years later) is able to discriminate between the fast and slow molecules of the gas in each of the two containers. The demon can open a small door between the two containers, letting only the fast molecules pass in one direction and only the slow molecules pass in the opposite direction. In this way the temperature in one of the two compartments would increase and in the other would decrease. This way heat flows in a direction (from cold to hot) which is forbidden by the second principle of thermodynamics. Equivalently, one could use this demon to reset the Carnot engine, that is to perform the transformation $B \rightarrow A$ without losing heat, obtaining a 100% efficiency (all heat transformed into work). In order to save the second principle and the evidence that 100%-efficient engines do not exist, one must conclude that such a demon spends energy to do his "discrimination" job. A task which, apparently, is mainly related to acquire information (measuring the speed of the molecules) has an inherent, irreducible, energy cost. We expand this subtle point in the entry Irreversibility.

Chapter 13
Fermi

Enrico Fermi (1901–1954) is recognized as one of the most important Italian scientists. Remarkably he has been one of the very few scientists who made fundamental contributions both to theoretical and experimental physics, in particular unraveling the secrets of nuclear interactions, and to the advancement of technology, with the creation of the world's first nuclear reactor, the Chicago Pile-1.

To really appreciate the significant role of Enrico Fermi (Fig. 13.1) in the physics of the twentieth century, it is useful to first recall the fascinating mystery of atoms (which is a much discussed topic in this book, see the entries Atoms, Boltzmann, Brownian Motion, and Einstein) going back to the years immediately before his scientific activity.

At the beginning of the twentieth century atoms were a solid hypothesis for a large part of the scientific community, though a tangible proof of their existence was still lacking and some important physicists were skeptical about it. The experiment of Perrin, published in 1908, based upon Einstein's theory of Brownian Motion, is considered a milestone that totally changed this state of things, making atoms indubitable. In Einstein's theory, atoms or molecules were very simple sources of random motion without any relevant internal structure. They could have been electrically neutral, not polarized, spheres that mechanically kick the wandering grain of pollen resulting in the beautiful and mysterious physics of Brownian Motion. But as soon as atoms became "true" elements of the physical reality, meaningful models of their structure started to appear.

Actually, the need to understand their internal structure emerged already a few years before Perrin experiment, since the discovery of the electrons by J. J. Thomson in 1897. Thomson considered the electron as a constituent of atoms. Being electrons negatively charged, in order to get electrically neutral atoms, some other part of atoms should have been positively charged, for compensation. Together with Lord Kelvin, Thomson believed that an atom was made of a positive "pudding" were negative "plums" were embedded. In 1909, just after Perrin demonstration of atom's existence, a series of experiment by Geiger and Marsden, interpreted by Rutherford

© Springer Nature Switzerland AG 2021
M. Cencini et al., *A Random Walk in Physics*,
https://doi.org/10.1007/978-3-030-72531-0_13

Fig. 13.1 Enrico Fermi, Courtesy of Archivio Amaldi, Dipartimento di Fisica, Università Sapienza, Roma

(all three worked in Manchester), gave convincing evidence for the existence of a positive nucleus surrounded by a cloud of negative electrons, similar to a system of planets orbiting around the Sun. In 1913, Niels Bohr (in Copenhagen) proposed a quantistic explanation of Rutherford model for the simplest case, the hydrogen atom (one electron orbiting around one proton): the angular momentum[1] of the orbiting electron can take only discrete values multiples of $\hbar$, the reduced Planck constant. Compatible with this idea, energy takes only values (called "levels" or "shells") in a discrete set. Energy released—as light—by electrons falling from one level to a lower one, can be emitted only in fixed amounts (*quanta*) and this explained the elusive atomic emission spectra observed in many experiments in the previous years. This was one of the first successful applications of the idea of quantized energy conjectured by Planck in 1900 to explain black-body radiation. Bohr extended his own atomic model to heavier atoms (more protons and more electrons) by adding rules for the occupation of the energy shells, resulting in a simple and suggestive

---

[1] A physical quantity linked to the rotational motion.

explanation for the mysterious chemical properties of all substances, codified by chemists of the nineteenth century in the periodic table of the elements.

In the years immediately following to the birth of Bohr's theory, Italy was involved in World War I and Enrico Fermi (born in Rome in 1901) was attending the high school, already revealing knowledge of mathematics and physics largely superior to that of his teachers. In 1918, he entered the University "Normale" of Pisa and immediately showed his talent. For instance, in 1920, he was invited to hold a conference, in front of his fellow students and professors, about the aforementioned theory of quanta, which was basically unknown in Italy. Before getting his degree in physics, in 1922, he already started publishing papers. Just after the degree, he visited Göttingen where he met Max Born, Werner Heisenberg, Wolfgang Pauli and Ernst Jordan, all among the chief founders of quantum mechanics, a theory which was being shaped exactly in those years to give a coherent meaning to all sparse quantum ideas of the two decades before (black-body radiation, photoelectric effect, Bohr's atom, etc.). In 1924 he went to Leiden to visit Paul Ehrenfest. Fermi was attracted to Ehrenfest's works on statistical physics (which Ehrenfest had learned by directly attending Boltzmann's lectures). Coming back to Italy, he took a teaching position at the University of Florence, between 1924 and 1926.

In 1926, Fermi published one of his most celebrated works, where he explained what happens if one has a system (such as a gas) made of many identical particles that obey the principle of exclusion—stated by Pauli just 1 year before. In a nutshell, the principle of exclusion affirms that only two electrons can be in the same orbital and they must have two opposite values of the spin number[2] ($\pm 1/2$). This principle complicates the probability of finding, in a gas of many electrons, one electron in a certain energetic state. To illustrate the issue inherent in the probability of finding electrons—not (or weakly) interacting—at a certain energy, one can imagine a game where balls (the electrons) are randomly thrown in cups (the energetic states). Boltzmann's statistics (see entries Boltzmann and Statistical Mechanics)—which disregards the exclusion principle—states that the cups are very large at small energy and very small at large energy: that is why one finds most easily classical particles at small energies. With the exclusion principle, things become more complicated because cups can contain only two electrons. Therefore, when throwing the balls the cups at smallest energies (more likely) are immediately filled, while only highest energies remain available and are progressively filled up. The probability distribution of balls/electrons—with this additional rule of the game—is quite different from Boltzmann's one, particularly at small temperature.[3] The idea of Fermi was independently developed in the same months by Paul Dirac and for this reason the formula for the occupation of energetic levels is now called Fermi-Dirac distribution. However, Dirac recognized that Fermi was the first to work on this problem and dubbed "fermions" the particles that obey the exclusion principle. In

---

[2]The spin of charged particles is a quantized feature of particles characterizing their interaction with a magnetic field.

[3]At small temperatures the tendency to occupy low energies is stronger, and this makes the exclusion principle more important.

the following years, it was understood that not only electrons, but all particles with a semi-integer spin number, belong to the family of fermions, which therefore includes also protons, neutrons, and many other sub-atomic particles. The particular shape of Fermi-Dirac distribution affects many properties of matter, for instance, several phenomena observed in metals, including the behavior of their electrical and heat conductivity.

In the fall of 1926, at age 25, Fermi got the (just created) position of Professor of Theoretical Physics at the University of Rome and started his very famous physics school with a small group of brilliant students and collaborators. The Physics Institute of the university was located in the center of Rome, in via Panisperna, and for this reason the group is widely known as "i ragazzi di via Panisperna" (the boys of via Panisperna). Fermi not only gave regular courses but also dedicated a large amount of time to solve problems *together* with his students (often just a few years younger than him), thinking aloud in front of them. In 10 years this method elevated the Italian school of Physics to a worldwide level and left in it an indelible sign. In the first years in Rome, Fermi dedicated his work to the applications of his statistical method to many aspects of the atomic structure of matter (see the entry Statistical Mechanics).

In 1932–1933, a series of fundamental discoveries about artificial radioactivity changed the knowledge of the physics of atom, in particular for what concerns the nature of its nucleus. Indeed 1932 is often considered the year of birth of nuclear physics. Fermi immediately intervened in this new field, publishing in 1933 a paper where he interpreted the nature of beta rays, a very penetrating radiation discovered in 1899 by Ernest Rutherford. Fermi's interpretation was correct[4] and remains as the basic scheme for the so-called beta decay even today: a beta ray is an electron emitted (together with an anti-neutrino[5]) from the spontaneous decay of a neutron into a proton. Such a spontaneous decay results from the fact that in nature there are substances whose atoms' nuclei are unstable and with the decay they reach a more stable energy level, leading also to a change of chemical species. The reason why there are unstable nucleus pertains to the high energy interactions between atoms in the core of stars or in Supernova explosions, where heavy nuclei with high energy levels are formed, some of which are unstable and, with their characteristic time, decay toward nuclei with lower energy levels.

A more important result obtained by Fermi in his work on nuclear physics is that of 1934, where he demonstrated the possibility to create radioactivity by shooting neutrons toward non-radioactive substances. The neutron is somehow captured by a stable nucleus and makes it unstable, i.e., the capture of the new neutron changes the energy profile of the nucleus. Fermi and his group started a rich experimental activity where they accumulated the knowledge of a sort of nuclear kitchen—not dissimilar to a scientific incarnation of magical alchemy—with a large series of recipes,

---

[4]Fermi initially submitted the paper which described his theory of beta decay to the influential journal Nature, which rejected it "*because it contained speculations too remote from reality to be of interest to the reader*". Years later Nature admitted that this was certainly one of its most serious mistakes.

[5]A neutrino is a very elusive elementary particle whose existence was hypothesized by Wolfgang Pauli in 1930 precisely to justify the non-conservation of energy in beta decay.

where neutrons were added to nuclei (like yeast to flour) to create new elements with the emission of various kinds of radiation. In the process of understanding this new kitchen handbook, Fermi discovered[6] an even more important phenomenon, namely, that the presence of water (or substances rich in hydrogen, like paraffin) hugely amplified the capture of neutrons in the nuclei and therefore their effectiveness in inducing artificial radioactivity. The reason is that hydrogen atoms slowed down the neutrons and made them more easily captured: such an explanation is quite counterintuitive (slowing down a bullet to make it more effective?) but appeared as a quite natural intuition in Fermi's genial mind, as he had a very solid knowledge of the essential laws of kinetic theory and statistical physics. It is fascinating to recall that the scattering—or diffusion—of neutrons through the target substance and the surrounding water is a phenomenon which is qualitatively very similar to the diffusion of the grain of pollen in the Brownian Motion experiment, and in fact Fermi investigated it theoretically on similar basis, using the concepts introduced by Boltzmann, Einstein and the great scientists who founded this fundamental subject (we suggest to see the entries Brownian Motion, Boltzmann, and Einstein). In the following years the quantitative understanding of the statistical phenomenon of neutron diffusion emerged to be crucial for all the most well-known applications of nuclear physics, such as nuclear reactors and nuclear bombs.

"For his demonstrations of the existence of new radioactive elements produced by neutron irradiation, and for his related discovery of nuclear reactions brought about by slow neutrons" Fermi received the Nobel prize in physics 1938. In the same year, the fascist Italian government started to persecute the Jews with a series of shameful racial laws, and Fermi decided to leave Italy as these laws menaced his wife's and his sons' life. Fermi after the Nobel prize ceremony reached Bohr in Copenhagen and from there (on 24 December 1938) with his family get on board the transatlantic directed to New York in the United States, where he was hosted at the Columbia University. There he immediately started to study uranium's fission (division): such a phenomenon, discovered a few months before by the German physicists Hahn and Strassmann, occurred in one of the nuclear reactions produced using Fermi's recipe. Fermi quickly realized that such a phenomenon—where a nucleus of uranium breaks down in smaller nuclei—should be accompanied by the emission of neutrons. And such an emission can be exploited to induce further fissions in surrounding nuclei, producing a possibly increasing chain reaction: one neutron breaks a nucleus, two (or more) neutrons are emitted, they break two (or more) nuclei, from these nuclei four (or more) neutrons are emitted, they break four (or more) nuclei, etc. Once this possibility was confirmed in few preliminary experiments, all research on this subject entered military secrecy. The reason is that the fission of uranium produces a large amount of energy and a chain reaction of such fission can release huge quantities of energy, capable of fueling machines, or destroying cities. Such a chain reaction was realized by Fermi with the first artificial nuclear reactor Chicago Pile-1 on the 2nd

---

[6]Here a quote of Fermi about what is a discovery: speaking about experiments *"There are two possible outcomes: if the result confirms the hypothesis, then you've made a measurement. If the result is contrary to the hypothesis, then you've made a discovery"*.

of December 1942 at the Chicago University. The entry into operation of Chicago Pile-1 is widely regarded as the moment when the era of nuclear energy began. The success of the Chicago Pile-1 reactor was reported in a coded phone call: *"The Italian navigator has just landed in the new world"*, and clearly, the "Italian navigator" was Fermi.

In 1944 Fermi moved to Los Alamos, together with some of the most important physicists in the world, in a secret laboratory-town created for the military applications of nuclear physics, where he testified the birth of the nuclear bomb. At the laboratories of Los Alamos, the selection of the best minds in the world discussing the solution of the difficult practical problems in the construction of the atomic bomb, led to the introduction of a whole series of new ideas that have revolutionized the course of science. Just to mention one of the most important computational method still in use, the Monte Carlo algorithm (see the entry Computer, algorithms, and simulations) was developed by Stanislaw Ulam in Los Alamos. However, the idea of Monte Carlo method, namely, the use of a statistical sampling technique to compute averages of physical quantities, was devised by Fermi many years before.[7] At Los Alamos, he also designed an analogical machine able to perform statistical computations of neutron transport and collisions called FERMIAC[8]

After the war, Fermi went back to the Chicago University, where the Institute for Nuclear Studies was created for him, though he preferred not to be the director to continue his nuclear physics research. In 1951 the building of a big cyclotron (a machine in which particles run in circles and then hit one against each other, breaking down and therefore emitting an entire zoo of different, and at that time often unknown, particles) allowed Fermi to enter the nascent field of sub-nuclear physics, focusing on mesons. During these researches he died, at the young age of 53, in 1954.

We find quite amazing that in only 50 years scientists went from the bare acceptance of the existence of atoms to the more and more detailed knowledge of their internal structure, including the ability to describe accurately the cloud of electrons (fundamental for chemistry) and penetrate the mysteries of the nucleus, learning to master nuclear reactions and radioactivity, reaching the incredible goal of breaking a

---

[7]*"Fermi took great delight in astonishing his Roman colleagues with his remarkably accurate,— too-good-to-believe —predictions of experimental results. After indulging himself, he revealed that his "guesses" were really derived from the statistical sampling techniques that he used to calculate with whenever insomnia struck in the wee morning hours! And so it was that nearly 15 years earlier, Fermi had independently developed the Monte Carlo method"*. From *The beginning of the Monte Carlo method* by N. Metropolis.

[8]*"The FERMIAC developed neutron genealogies in two dimensions, that is, in a plane, by generating the site of the "next collision". Each generation was based on a choice of parameters that characterized the particular material being traversed. When a material boundary was crossed, another choice was made appropriate to the new material. The device could accommodate two neutron energies, referred to as "slow" and "fast". Once again, the Master had just the right feel for what was meaningful and relevant to do in the pursuit of science"*. From *The beginning of the Monte Carlo method* by N. Metropolis. This is a shining example of how deep physical insight can simplify a complex problem and make it manageable, see entry Laws, levels of description and models.

nucleus and extracting an unimaginable energy from it (unfortunately used also for military purposes), and finally start to see the internal structure of the tiny particles forming these nuclei. It is equally—or even more—amazing to realize that a large part of this journey had among his protagonists a single man, Enrico Fermi, who in his life was one of the few physicists in the twentiethcentury to fully conjugate theory and experiments. We cannot avoid to highlight that in this impressive journey of human intellect, Fermi (as many of the scientists working in atomic and nuclear physics) made frequently use of the tools of statistical physics and kinetic theory to rationalize experimental discoveries: it was impossible (in fact it is still today very difficult) to visualize single atoms and/or their internal structure, so that every observation was indirect and mediated by a large amount of events, so that only statistical information was available. We can safely affirm that the development of quantum theory is strictly linked to statistical mechanics.

The study of statistical physics was constantly present in Fermi's work, particularly in its first part, when his interest was for electrons and their interaction with nuclei. One of the most famous legacies left by Enrico Fermi to physics is the so-called Thomas-Fermi model, where he (independently, a few months after L. H. Thomas) modeled the cloud of electrons as a continuum density and from this gave a first approximation of an atom's energy based upon simple statistical arguments. These arguments are the basis of the modern density functional theory, widely used in solid state physics.

The importance of understanding the foundations of statistical physics (and even pure probability) was clear to Fermi who, in fact, not only used its methods for his discoveries in atomic/nuclear physics, but also wrote papers focused only on this subject. The 1926 paper on the statistics of a gas of fermions is, of course, his most famous. There are, however, other less widely known papers which testify his vivid interest for the most fundamental aspects of statistical mechanics. A theme of research which appears in several works by Fermi is that of *ergodicity* which is a concept underlying statistical mechanics, see the related entry. For instance, in 1923, he wrote an interesting paper with a partial demonstration of ergodicity, by making a few assumptions which—in the following years—were understood to be too restrictive, i.e., basically not true in many systems (see also the entry Kolmogorov). Fermi remained deeply interested in the problem of ergodicity and continued thinking and working on it until his death. In fact in 1954 he realized, together with J. Pasta, S. Ulam , and M. Tsingou, one of the first numerical experiments of history. The use of computers started a few years before, basically triggered by the need of huge computations for military purposes, in particular for the development of nuclear bombs (see the entries Von Neumann and Computer, algorithms and simulations), and for this reason was a tool that Fermi considered very interesting and helpful since its birth. Fermi, Pasta, and Ulam[9] used their computer ("MANIAC" was its name) to

---

[9]This pioneering paper is typically also known as "FPU": actually another author, Mary Tsingou, worked in programming the MANIAC computer.

calculate the trajectory of an ideal mechanical system,[10] looking for the evidence of ergodicity. The result of the computation was quite surprising: the computer produced many trajectories which are not at all ergodic. Roughly speaking, this corresponds to a situation where the "billiard balls" run across the table in a very regular periodic fashion (see the entry Statistical mechanics), exploring only certain special paths, without filling the whole space. This surprising result in the very same years was understood theoretically by Kolmogorov, Arnold and Moser who—independently—contributed to what is now known as "KAM" theorem (from the initials of the three names). This work by Fermi and co-workers is considered as a first example of numerical *gedankenexperiment* (ideal experiment): until that moment the computer had been used to obtain the results of complicate computations, while in this case it had been used to verify a theoretical hypothesis. From that moment this way of using computers in physics has become more and more widespread. Today it is a fundamental part of research in theoretical physics. Fermi, also in that aspect, demonstrated to be a precursor of modern times.

---

[10]Technically Fermi, Pasta, and Ulam simulated a Hamiltonian system of oscillators coupled by harmonic and weakly anharmonic interactions.

Chapter 14
Fractals

Order and harmony have always played a central role in the mathematical tradition, it is thus not surprising that regular geometrical objects (think of the Platonic solids)[1] and regular functions (e.g., differentiable, or at least differentiable excepts for some isolated points) have been the primary focus of mathematical investigations. We can mention a celebrated passage of Galileo, which can be considered as the manifesto of the mathematical program to describe the physical reality: *"Philosophy is written in this grand book—I mean the universe—which stands continually open to our gaze, but it cannot be understood unless one first learns to comprehend the language and interpret the characters in which it is written. It is written in the language of mathematics, and its characters are triangles, circles, and other geometrical figures, without which it is humanly impossible to understand a single word of it; without these, one is wandering around in a dark labyrinth"*. On the other hand, in the words of Benoit Mandelbrot, it is well evident that *"Clouds are not spheres, mountains are not cones, coastlines are not circles, and bark is not smooth, nor does lightning travel in a straight line"*. Therefore, while acknowledging the absolute importance of classical geometry, the above examples (to which many more could be added) clearly demonstrate that to describe nature, in many circumstances, one needs to go beyond the old good classical mathematics based on regular geometry and smooth functions.

From the second half of the nineteenth century to the first half of the twentieth century only a few scientists had the intuition of the importance of "irregular"

---

[1]The tetrahedron, the cube, the octahedron, dodecahedron, and icosahedron with, respectively, 4, 6, 8, 12, and 20 faces played a prominent role in Plato's philosophy. In his *Timaeus*, he associated to four of them the four classical elements (earth, air, water, and fire). About the dodecahedron he wrote *"the god used [it] for arranging the constellations on the whole heaven"*. They were mathematically described by Euclid, who argued that they were the only regular (with equal faces) polyhedra. To appreciate their influence in philosophy and science think that Kepler proposed a model of the Solar System based on the platonic solids, before surrendering to the factual evidence that the correct geometry grounds on ellipses.

© Springer Nature Switzerland AG 2021
M. Cencini et al., *A Random Walk in Physics*,
https://doi.org/10.1007/978-3-030-72531-0_14

mathematical objects. Among such exceptions we can mention Weierstrass, Peano, Hausdorff, and Julia with their studies of "mathematical monsters" and Perrin and Richardson with theirs works on the Brownian Motion and turbulence. However, the man who fully realized and disseminated, also at a popular level, the widespread presence and importance of such irregular objects in natural sciences has been Benoit Mandelbrot (1924–2010) who coined the term "fractal" (for reasons that will be clear soon). Since the 1960s, inspired by the contributions of the previously mentioned precursors, Mandelbrot developed a new geometry—the fractal geometry—, now well known even among the general public. Possibly, such a widespread success is mainly due to the visual beauty and aesthetical aspects of many fractal objects, however, it is important to stress that the relevance of fractal geometry goes much beyond aesthetics.

Due to his importance, we briefly outline Mandelbrot biography. He had Polish-Lithuan origin and studied at the École Polytechnique in Paris. After World War II, French mathematics was dominated by the Bourbaki group characterized by a very formal approach. Mandelbrot was not sympathetic at all with that academic climate and moved to the United States, where he earned a master's degree in aeronautics. Then, he obtained his Ph.D. in mathematics in Paris and, after short periods in France and Switzerland, in 1957 finally moved back to US with an appointment at IBM and Yale University. During his long scientific life Mandelbrot worked in many different fields: information theory, finance, thermodynamics, linguistic, turbulence, cosmology, geophysics, and more. He had professorships rather disparate as economy, applied mathematics, engineering, and physiology. However, he is mostly known for his studies on fractals and self-similarity.

To exemplify what a fractal is and the origin of this name, it is worth examining with some details a specific case, namely, the well known von Koch curve. This geometrical object is built as follows. Take a segment of unit length and divide it in three equal parts. Remove the central part and substitute it with two segments of the same size (forming an equilateral triangle with the bottom side removed), so we have a curve with 4 segments of size $1/3$. Repeating such a procedure n times on each segment, we obtain 4^n segments of size 3^{-n}, so that the total length of the curve is $4^n \times 3^{-n} = (4/3)^n$, which diverges as $n \to \infty$. The curve (and the procedure to derive it) is illustrated in Fig. 14.1, where one can also appreciate the self-similarity typical of fractal objects, namely, the fact that a small portion of the curve reproduces, on a smaller scale, the whole curve (see also the Weisestrass curve in the entry Richardson), and the origin of the name fractal, coming from the Latin *fractus* meaning interrupted, irregular, literally broken.

To characterize such a curve it is useful to introduce the concept of fractal dimension. In order to do that it is useful some preliminarily considerations. We say that a regular curve has dimension 1 and the surface of a sphere has dimension 2, because in the former case one variable (the curvilinear coordinate) is sufficient to determine a point on it, while two coordinates (latitude and longitude) are used in the latter. On the other hand, we can introduce the concept of dimension with a different reasoning. We can approximate the curve with a broken line of segments of size ℓ and count the number of segments, $N(\ell)$, necessary to approximate the curve. The broken line

Fig. 14.1 Representation of
a few steps of the
construction of the von Kock
curve

has length $L(\ell) = N(\ell)\ell$. To obtain the length of the curve, $\mathcal{L}$, we have to repeat
such a procedure taking ℓ smaller and smaller, i.e., taking the limit of vanishing ℓ.
In the case of a regular curve, in such a limit, $N(\ell)\ell$ will approach the curve length
$\mathcal{L}$, i.e., $\lim_{\ell \to 0} N(\ell)\ell = \mathcal{L}$. The existence of such a limit implies that $N(\ell) \sim \ell^{-1}$
for very small ℓ. Similarly, we can "tile" a regular surface with $N(\ell)$ small squares
of size ℓ, and the area $\mathcal{A}$ is well approximated by $N(\ell)\ell^2$, therefore $N(\ell) \sim \ell^{-2}$.
Repeating the procedure with a regular three-dimensional object we easily realize
that $N(\ell) \sim \ell^{-3}$. Therefore the dimension of a geometrical "standard" objects can
also be obtained by looking at the scaling exponent of $N(\ell)$ for very small ℓ, i.e.,
$N(\ell) \sim \ell^{-D}$.

What if we perform the above procedure to measure the length of the von Kock
curve? We already demonstrated that its length is infinite, what does it means in
terms of the above procedure to define the dimension? For the von Kock curve,
we can use $\ell = 3^{-n}$ and we clearly obtain $N(\ell) = 4^n$, which we can rewrite[2] as
$N(\ell) = \ell^{-\ln 4 / \ln 3}$ implying a non-integer—fractal—dimension $D_f = \ln 4 / \ln 3 \simeq$
1.2618. This result, which is specific for the von Kock curve, is actually quite general
as it turns out that most of the irregular objects quoted above can be described in
terms of a non-integer (fractal) dimension.

More important is to understand whether such kinds of objects are just mathematical pathologies or are physically relevant. In the early twentieth century, Perrin

---

[2]Using the function logarithm in the relations $\ell = 3^{-n}$ and $N(\ell) = 4^n$ it is possible to rewrite them
as $\ln \ell = -n \ln 3$ and $\ln N = n \ln 4$. Using the first expression one obtains n as a function of ℓ,
i.e., $n = -\ln \ell / \ln 3$, and substituting it in the second espression gives $\ln N = -(\ln \ell) \ln 4 / \ln 3 =$
$\ln \ell^{-\ln 4 / \ln 3}$ and finally $N(\ell) = \ell^{-\ln 4 / \ln 3}$. Let us stress that in this particular example the use of
$\ell = 3^n$ makes not necessary to take the limit $\ell \to 0$. A different choice for ℓ, however, leads to the
same result, provided ℓ is small enough.

had been among the first one to realize the relevance of self-similarity: *"Consider, for instance, one of the white flakes that are obtained by salting a soap solution. At a distance its contour may appear sharply defined, but as soon as we draw nearer its sharpness disappears. [...] The use of magnifying glass or microscope leaves us just as uncertain, for every time we increase the magnification we find fresh irregularities appearing, and we never succeed in getting a sharp, smooth impression, such as that given, for example, by a steel ball"*. The first to ask the apparently silly question *How long is the coast of Britain?* was not Mandelbrot (who is usually regarded as the father of fractals), but the English scientist L. F. Richardson. In a posthumous paper he plotted as a function of the resolution ℓ the length $L(\ell)$ of the coastlines of Great Britain, of the land border of Germany, of the Spain-Portugal border, and of the coastlines of Australia and South Africa. In a regular curve for small ℓ one has that $L(\ell)$ approaches a constant, on the contrary, Richardson observed a behavior of the form $L(\ell) \sim \ell^{-\alpha}$ where α is about zero for the coastline of South Africa, while in the other cases it is positive, and increases with the "wrinkleness" of the curve. In terms of fractal geometry the link between α and D_f previously defined is $\alpha = D_f - 1$. Let us note that this behavior, observed in an empirical way, is all alike that of the von Koch curve. As discussed in the entry Richardson, this very original scientist was one of the first to discuss the idea of self-similarity which had played a key role in science starting from 1960s, see also the entries Turbulence and Universality.

One of the first physical examples of self-similarity dates back to the discovery of Brownian Motion by the botanist Robert Brown (1827) who was observing, with a microscope, a grain of pollen suspended in water. By looking at the irregular motion of one of the spatial component, $x(t)$, of a Brownian particle, at long time one has that the mean square displacement $\langle [x(t) - x(0)]^2 \rangle \simeq 2Dt$, where $\langle \, \rangle$ denotes the average (for instance on many particles) and D the diffusion coefficient, which Einstein linked to the Avogadro number.[3] The above relation between the square displacement and time does suggest that the curve cannot be differentiable. Indeed, a few years after the works of Einstein and Perrin, in the 1920s, Wiener introduced a mathematical formalization of the Brownian Motion in terms of the now called Wiener process, that is a stochastic Gaussian process, continuous but everywhere not differentiable such that $\Delta x(\Delta t) = x(t + \Delta t) - x(t) \sim \sqrt{\Delta t}$. In a more formal way, the variable $\Delta x(\lambda \Delta t) = x(t + \lambda \Delta t) - x(t)$ has the same properties of $\sqrt{\lambda} \Delta x(\Delta t)$, i.e., the Wiener process has a self-similarity structure: rescaling the time t by a factor λ, and the x by a factor $\sqrt{\lambda}$ we have the same (statistical) features. Applying to a graphical representation of a Wiener process (a one-dimensional Brownian Motion), i.e., a graph of $x(t)$ versus t, the procedure to compute the length at varying Δt, one obtains $D_F = 3/2$, again an example of fractal dimension.

Since its recognition fractal geometry stimulated the development of new mathematical ideas and methods, with an interesting cross-fertilization with physics, probability, and information theory, which positively affected our understanding of a wide class of phenomena including turbulent flows, earthquake dynamics, ecosystems,

---

[3] As discussed in the entries Brownian Motion and Atoms, the study of such an irregular wandering of small particles in water led to the confirmation of the atomistic hypothesis.

large scale structure of galaxies, percolation, and geophysical applications. In the following we present a non-exhaustive list of topics where the concepts of fractal geometry had been useful.

Chaos: Fractals naturally appear in chaotic dynamical systems.[4] In particular, when the system is dissipative[5] its long time dynamics evolve onto a so-called attractor[6] which can be a point, a smooth curve (limit cycle) or a complex object, in the latter case we speak about a *strange attractor*, a term originally due to David Ruelle. Such strange attractors appear whenever the system is chaotic, that is when there is sensitive dependence on initial conditions (i.e., when a small error on the initial condition is exponentially amplified). As shown by Ruelle, such strange attractors cannot be regular geometrical objects and typically display self-similar, fractal properties. An intuitive way to characterize such fractal structures is to look at the fraction $M(\ell)$ of points belonging to the attractor in a sphere of radius ℓ centered on a point of the attractor, we have $M(\ell) \sim \ell^{D_f}$. Just to give the idea: the celebrated Lorenz model is characterized by a strange attractor that while living in a three-dimensional space has a non-integer, fractal dimension about 2.05. Actually, rather often the scenario is more complex and a single dimension is not enough to characterize the attractors. In such cases we speak of multifractals, which loosely speaking means that different points of the attractor are characterized by different fractal dimensions. From this short survey, it should be clear that fractal geometry played a key role in the understanding of dissipative chaotic systems. For more information see the entry Chaos.

Turbulence: It is commonly experienced when flying on an airplane to encounter turbulent spots which make the plane to vibrate even very violently. Such vibrations are due to the very disordered and intense motion of the surrounding air. Turbulence is a quite ubiquitous phenomenon happening whenever the Reynolds number (a non dimensional number measuring the strength of the non-linear forces acting on the fluid over the linear ones) becomes very large. It is characterized by an irregular velocity wildly varying both in time and space (spatiotemporal disorder). As shown by Kolmogorov in 1941, in a turbulent velocity field, in an appropriate range of scales (not too small and not too large), the difference of the velocities between two points at distance ℓ varies approximately as $\ell^{1/3}$, i.e., $\Delta v(\ell) \sim \ell^{1/3}$. Such a scaling behavior is conceptually very similar to the above discussed Wiener process

---

[4]The term dynamical systems typically designates those system whose evolution is described by (time continuous) differential equations or (time discrete) maps. Mathematically the former can be formalized as $dx/dt = F(x)$, and the latter as $x(t+1) = F(x(t-1))$ where the vector $x(t)$ represents the state variables describing the system, e.g., position and velocities of particles, and F is the evolution law, typically a non-linear, differentiable function.

[5]For instance, an example of dissipative system is the damped pendulum, which is described by the equations $dx_1/dt = x_2$, $dx_2 = -\gamma x_2 - g/L \sin x_1$, where $x = (x_1, x_2)$ are the angle and angular velocity, respectively, and $-\gamma x_2$ represents the air friction.

[6]Again thinking of the damped pendulum, one can realize, also referring to daily experience, that at long time it will end in the state $(x_1, x_2) = (0, 0)$ meaning the pendulum at rest vertically oriented. This state is called attractor for the dynamics (since it attracts all the possible trajectories). All dissipative systems are characterized by an attractor.

where $\Delta x(\Delta t) \sim \Delta t^{1/2}$. As discussed in the entry Turbulence, the velocity field in a turbulent flow can be characterized by the fractal (actually multifractal) formalism and has a measurable impact on the behavior of small particles transported by the flow. Indeed the evolution of the distance between two particles depends on the velocity difference over their separation. The above scaling behavior implies an explosive (extremely fast) separation between the particles: the average square particle distance grows as t^3 in a time t. We can surely say that the current understanding of the statistical properties of turbulent flow is deeply influenced by fractal geometry.

Astrophysics: Let us briefly recall an old paradox introduced by Olbers (nineteenth century): in a static universe whose large scale mass distribution is uniform, the intensity of light must be necessarily infinite. Now astronomers believe that the solution of the paradox lies in the expansion of the universe. Nevertheless, interestingly, the paradox can be explained also in a static universe if the star distribution is fractal with $D_f < 2$.[7] The French astrophysicist de Vaucoulers in the 1970s, from accurate data analysis, derived that the density of matter within a sphere of radius R is proportional to $R^{-\alpha}$ with $\alpha \simeq 1.8$, in terms of the fractal geometry this corresponds to a fractal distribution of the galaxies with dimension $D_f = 3 - \alpha \simeq 1.2$. Recent data on rather large and accurate catalogues show the same qualitative features with $D_f \simeq 2$. There is an agreement among the experts that this fractal structure holds at least up to $R \sim 20$ megaparsec (i.e., about 6.5×10^7 light years).

Medicine: The blood vessels and the bronchial trees of mammals show a selfsimilar fractal structure. Some scientists conjectured that such a structure is the optimal one for the most efficient distribution of blood in the arteries and of oxygen in the bronchi. Fractal models of blood vessels are systematically used in numerical computation for haemodynamics.

Geophysics: In the dynamics of the atmosphere and of the oceans turbulent phenomena play a key role. For a realistic description it is necessary to take into account several aspects as the dynamics of the temperature, the Coriolis effect, and so on, therefore the scenario à la Kolmogorov is typically too simplified to describe such complex dynamics, however self-similar structures are observed and typically one has (multi)fractal distributions. Fractal structures are also present in seismic faults, coastlines, and river deltas. It is pretty impossible to build a theoretical description of such complex systems from first principles; fortunately, sometimes, with the use of simplified fractal models it is possible to reproduce the main features of the observed data. For instance, this happens in the modeling of the formation of the coastlines.

---

[7]In a universe with a uniform distribution of stars, at a distance between R and $R + dR$ from a given point there are a number of stars proportional to $R^2 dR$, so the contribution is proportional to dR; since the luminosity is proportional to R^{-2}, after integrating we get an infinite value for the luminosity. It is easy to repeat the computation for a distribution with fractal dimension D_f: now at a distance between R and $R + dR$ from a star there are a number of stars proportional to $R^{D_f - 1} dR$, since the luminosity is proportional to R^{-2}, the contribution is proportional to $R^{D_f - 3} dR$ therefore for $D_f < 2$ the integral converges.

Fractals at Hollywood: Let us conclude with an application of fractals in the cinema industry. Many artificial images of films, in particular science-fiction and fantasy ones, are obtained with a mathematical technique (called iterated function system) able to produce pictures with a self-similar structures. Examples of such nice computed-generated images are landscapes, fantasy architectures, clouds, aurorae, and fires.

Chapter 15
Information Theory

"W ar sre tht u wil nt hv ay difcty i udstdg th information cnted i tis sntce altogh w rmovd a lrge nmbr o letrs".[1] It might have taken a few more seconds that if we had used all the letters. But if you need to send the above sentence between quotes to a friend and this costs you a dollar per letter you would have spared 44 dollars sending the shortened message: a good deal if your friend does not get nervous.[2] You would have spared a few more contracting *information*, e.g., in info. We have not done it, because information is precisely the subject of this short discussion. On the other hand, it is easy to realize that you cannot spare much money if you need to send the following message: "njkncnwdjknvjyken dvn dkfv ljkdfnkjvsndfvnodfvkrndsfjkn-vkjldsn vjky ncsjkcv kzjxacnvkljsbndl baxhj baszi nbobvjajscbxalsnc asdgywedfn ojnocvsn jiadjswc dqsf". What is the difference between the two messages? They are both using the same alphabet (27 symbols: 26 letters plus space), their length is the same (155 symbols, excluding the full stop), but yet they are different. We can recover the first one using only 111 characters, i.e., eliminating about one every three characters. While the same does not seem to be possible in the second one. Why? The easy answer could be the first is written in English while the other is either in a different (unknown) language or is just a random concatenation of letters. In both cases we cannot *compress* its information. Intuitively we understand that (without consideration of the meaning) the second message contains more information and thus it is harder to compress. Indeed it is easier to complete "udstdg" in "understanding" as we know of many instances in which 'u' and 'd' are separated by 'n' and followed by 'er'. While, for instance, in the second message 'k' is followed by many possible letters independently of the previous or next one.

The possibility to compress the first message is due to the fact that English (and this is generically true in all languages) is redundant. Even without knowing English,

---

[1] Just to be sure, we repeat: "We are sure that you will not have any difficulty in understanding the information contained in this sentence although we removed a large number of letters."

[2] Eliminating only the spaces, another possibility, you would have spared only 25 dollars.

© Springer Nature Switzerland AG 2021

M. Cencini et al., *A Random Walk in Physics*,

https://doi.org/10.1007/978-3-030-72531-0_15

if you are given a collection of English texts you would be able to reconstruct (a large part of the words of) the first (contracted) message by studying the frequencies of occurrence of letters given the preceding ones. Of course, knowing English it is even easier because you are helped by the meaning of what you are reading. The first option is interesting as it alludes to the possibility to quantify the information of the message regardless of its meaning. This is precisely the goal of information theory.

Information theory is one of those branches of science which had the privilege of being born mature, when formulated by Claude Shannon and his colleague Warren Weaver in 1948. It has been refined and expanded in the successive years, but all the ingredients were essentially present in the original formulation. It was born for a very practical problem, that is to communicate a message through a device (communication channel) which, for several reasons, could be noisy and thus can contaminate the message. For instance, if we had to communicate the starting sentence through a device which eliminates one letter every three (or changes it randomly with another character) we would have avoided to make it shorter. In other terms we would have exploited the redundancy of English language to overcome the errors due to the communication channel.

In the following we shall provide some (hopefully simple) hints on how these ideas can be formalized. To this aim we shall need some abstraction, and using (discrete) numbers instead of letters can help. A message can be anything emitted by any type of source. If the source is a book the message will be the sequences of letters (which can be converted in numbers, e.g., the ASCII code), if it is two people playing coin tossing it will be head/tails (0 and 1), if it is a football matches is win/lose/draw (e.g., 1/-1/0), a film on TV it will be a sequence of images each of which can be digitized (again transformed in numbers), etc. As we shall see later, the source can also be the trajectory of a chaotic system (see the entry Chaos).

Information theory does not deal with the single message but characterizes the information content of the ensemble of messages emitted by a source. Thus it quantifies the (average) information produced by the source. In particular, when the source is ergodic (see the entry Statistical mechanics for a discussion on ergodicity) the average can be performed over a single (very long) message instead of an ensemble of messages. The question now is: How to measure the information of the source emitting the messages?

In order to grasp the basic idea, consider the following example. Assume that the source is a urn containing 8 balls with numbers from 0 to 7, and that every now and then a number is extracted randomly and independently from previous extractions and, after recording it, is put again in the urn so that each number has a probability 1/8 to be picked up. What is the information we gain knowing the outcome of each extraction? Intuitively, you can quantify the information of each extraction by counting the number of questions you have to ask to know that number. It is quite easy to realize that you just need 3 questions. Say the number is 6. You may ask whether the outcome is larger than 3? Yes, then you ask if it is smaller than 5? No, then it can only be 6 or 7, which will be your last question.[3] The probability of each number is $p = 1/8 = 2^{-3}$ and you just needed $n = \log_2(1/p) = -\log_2 p = 3$ questions. For equiprobable outcomes $1/p$ is just the number of possibilities. As discussed in the entry Entropy of this book, the log of the number of possibilities is called entropy in physics, and the same name is used in information theory.[4]

Is the inverse probability still good when the outcomes are not equiprobable? Less frequent events surprise you more than more probable ones, i.e., they carry more information, so the use of the inverse of the probability agrees with intuition. You can also generalize the above example, assuming a difference in the probability of extraction of the numbers, e.g., one of the numbers is more probable than the others. Well in this case you can convince yourself that, on average, you need less than 3 questions (hint: your first question should be whether the outcome is the most probable number). Repeating the game a large number of times you will find that on average the number of questions to discover the outcome of an extraction is (with the convention that $0 \log_2 0 = 0$, also recall that probabilities should sum to 1)

$$h_{Sh} = \sum_{i=1}^{M} p_i \log_2(1/p_i) = - \sum_{i=1}^{M} p_i \log_2 p_i$$

that is the Shannon entropy of the source, i.e., the information of the outcome i $(-\log_2 p_i)$ averaged over all the possible ones, $i = 1, \ldots, M$, according to their probability of occurrence p_i. Notice that if all probabilities are equal $p_i = p = 1/M$ we have $h_{Sh} = -\log_2 p = \log_2 M$, which is precisely what we derived in the example with the eight numbers. If the source can emit only a certain number, say 1, i.e., $p_1 = 1$, and thus all the others are impossible ($p_i = 0$ for $i \neq 1$) then the entropy is zero, and indeed we get no information as the emitted outcome 1 is certain. Finally, using some simple mathematics (namely, using the convexity of logarithms), one can show that, in general, $h_{Sh} \leq \log_2 M$, and this is also quite desirable as the most uncertain (more informative) situation is when all outcomes happen with the same probability $1/M$.

The use of the logarithm is also crucial if we demand the natural property that guessing two numbers requires twice the number of questions than guessing one. This is so because the probability of a sequence of two numbers is given by the product the probabilities of each number, i.e., if i and j have probability p_i and p_j, the probability to observe ij is $p_{ij} = p_i p_j$. Since $\log_2(p_i p_j) = \log_2 p_i + \log_2 p_j$, one can easily generalize the above definition of Shannon entropy to find that the Shannon entropy for K extractions is just $H_K = K h_{Sh}$. This is called the additivity of entropy. Additivity, however, is not generic as $p_{ij} = p_i p_j$ holds only if i

---

[3]This procedure is an example of binary search tree, much used in computer science. For instance, it allows to sort M numbers with $\log_2 M$ comparisons between pairs of numbers.

[4]In information theory usually one uses the logarithm base 2, as it corresponds to measuring information in number of bits, i.e., the length of the string of 0 and 1 used to code a given number or letter (e.g., think of the ASCII code). For instance, a three bits encoding of 5 is 101, i.e., $5 = 1 \cdot 2^0 + 0 \cdot 2^1 + 1 \cdot 2^2$.

and j are independent. For instance, it may be that extracting 2 is more probable if the previous number was 5, so reading 2 after 5 is less informative, like if we see a "u" after a "q" in an English word. In general, since knowing more of the past cannot increase the uncertainty on the future, we should expect entropy to be sub-additive which mathematically can be expressed by saying that if H_K is the information content of a sequence of K symbols—usually dubbed K-word—, then $H_{K+1} \leq H_K + H_1$ (i.e., the information content, the entropy, of $(K + 1)$-words will be no larger than the sum of the entropy of K-words and that of 1-words) though of course $H_{K+1} \geq H_K$. Being more quantitative would require to introduce joint and conditional probabilities, which we prefer to avoid here. For a generic source, emitting non-independent symbols, the entropy per symbol can be defined as follows. Consider long sequences of K symbols, compute their entropy, H_K, from the probability of occurrence of the K-words. Then the Shannon entropy per symbol is obtained as the limit $h_{Sh} = \lim_{K \to \infty} H_K/K$. The existence of the limit implies that the entropy of K-words asymptotically (for large K) behaves as $H_K = K h_{Sh}$ (plus something which can be neglected for large K). Intuitively this means that, for large K, the number of possible K-words actually observed is $M = 2^{K h_{Sh}}$ and all of them are equiprobable with probability $1/M = 2^{-K h_{Sh}}$, from which the link with the statistical mechanics interpretation of thermodynamics entropy (see the entry Entropy) should be clear. This result can be made rigorous and goes under the name of Shannon-McMillan-Breinam or asymptotic equiprobability theorem.

The problem of coding messages, which is crucial in communication (for which information theory was originally developed), offers another intuitive view of entropy. Assume you want to transmit a certain number of words (sequences of symbols) in the less expensive way, i.e., the most compressed one (this is what you typically do while zipping large attachments in your emails). How should you code the words, e.g., in bits (sequences of 0 and 1), for the average length of the code to be the shorter one? We illustrate the basic idea with the following example. For simplicity, say that we have three words (w_1, w_2, w_3) of equal length appearing in the message with probabilities $(p_1, p_2, p_3) = (1/2, 1/4, 1/4)$. If we code them with a string of symbols of equal length ℓ, e.g., with $\ell = 3$ we can code the words with the following binary strings 001, 010, 100. This is not very efficient, as the w_1 occurs twice more frequently than each of the other two. So it makes sense to use for w_1 a shorter string with respect to the other two. According to Shannon an optimal choice would be to encode a word w_i with $\ell_i = -\log_2 p_i$ bits. In the above case, a possibility is $w_1 \to 0$, $w_2 \to 10$, $w_3 \to 11$. In this way the average length would precisely be $\langle \ell \rangle = \sum_i p_i \ell_i = h_{Sh} = 1.5$ which is half the naive coding. While specialized for this numeric example this idea can be easily generalized. Actually this strategy is not always optimal and in general we have $\langle \ell \rangle \geq h_{Sh}$,[5] i.e., the Shannon entropy sets a limit to our possibility to compress a message.

Back to the English example with which we started, we can now understand that the entropy of English (with its alphabet of 27 symbols, letters plus space) should be smaller than $\log_2 27 = 4.76$, and that is why we could have used just

---

[5]Huffman in 1952 introduced a coding strategy which is essentially optimal.

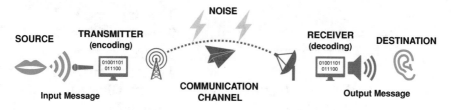

Fig. 15.1 Sketch of the processes and elements involved in communication

111 characters instead of 155 (in this case the coding was done using the same alphabet). Actually, our exercise suggests that the entropy of English is smaller than $111/155 \log_2 27 = 3.4\ldots$. Indeed, Shannon estimated the entropy of English to be around 2.62 and obtained such results, asking people to guess letters in a text given the previous ones. In general, if messages are coded with the same alphabet of M symbols and the entropy h_{Sh} is less that its maximum allowed value $\log_2 M$, no coding strategy can compress the length of messages by more than a factor $h_{Sh}/\log_2 M$, and we can define the redundancy of the source emitting the messages as $1 - h_{Sh}/\log_2 M$.

We wish now to discuss two applications of information theory. The first one consists in two simple working examples of communication, which are expedient to see the potentiality of the approach for the scope it was originally conceived and to introduce the important concept of mutual information. The second one is an application of information theory to chaotic systems.

The processes involved in communications are schematically shown in Fig. 15.1. So far we discussed the first element, namely, the *source*, and we characterized it quantifying the information per symbol emitted in terms of the Shannon entropy. The *transmitter* acts on the signal emitted by the source coding it (e.g., in strings of bits) in a way suitable for being transmitted, we have already given some simple idea of coding. Then we have the *communication channel*, which in real applications will always contaminate the transmitted messages due to the presence of some noise. Finally we have the *receiver* where the message is decoded and made available to the *destination*. The properties of the channel are the most delicate ones, as the limits to the rate at which we can transmit the messages is basically set by the channel properties, in particular by the *channel capacity* which, as we will see, is again related to entropy. Schematically a communication channel takes as input a symbol x_{in} (for us a random variable, e.g., 0 or 1) deriving from a source, and delivers at the end an output symbol x_{out} (again a random variable) that depends probabilistically on the input. Clearly, if the output is completely independent from the input (e.g., $x_{in} = 0$ and $x_{out} = 1$ or 0 with probability $1/2$) no communication is possible, meaning that the channel capacity is 0. While, if x_{out} is completely determined by x_{in} it simply means that the channel is noiseless. For instance, for a binary noiseless channel, shown in Fig. 15.2a, if $x_{in} = 0$ or 1 then $x_{out} = 0$ or 1, respectively with certainty, in this case the channel capacity is maximal and equal to 1 bit. Consider now the channel sketched in Fig. 15.2b, it allows in principle for a 2 bits transmission as it uses four symbols. However, due to noise each input symbol will be received with

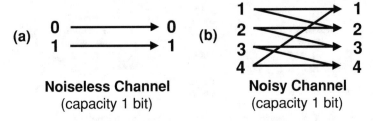

Fig. 15.2 Two simple examples of communication channels

probability $1/2$ either as the same symbol or as the next symbol as shown in the figure. In this case it is easy to understand that if we use only symbols 1 and 3 to code the message we can reconstruct it with no ambiguity (as, e.g., receiving 1 or 2 means that the input was 1). This means that the noisy four symbols channel is equivalent to the noiseless binary channel and thus it has capacity equal to 1 bit (2 symbols). In general, intuition suggests that we can communicate if the knowledge of the output reduces the uncertainty on the input. A measure of such reduction is provided by the mutual information, $I(X_{in}, X_{out})$ between the input/output variables (the use of capital letters in the argument of I is to denote the random variables and not their specific values), which measures the amount of information on the input obtained observing the output.[6] We add that the actual channel capacity is given by the supremum of the mutual information over all possible probability distribution of the inputs, as the channel can serve different sources. Back to the last example, computing the channel capacity would reveal that it is indeed 1 bit as obtained by construction. We emphasize that, in general, communication channels are not as simple as the examples examined above, and it is not always clear how to identify a way to send the messages without errors. However, there are strategies to reduce the problem to a similar situation.

We now move to the second application, that is how to use information theory and, in particular, entropy to characterize chaotic systems. As discussed in the entry Chaos of this book, the hallmarks of chaotic behaviors are the so-called sensitive dependence on initial conditions, that is the exponential growth (controlled by the Lyapunov exponents) of the distance between two trajectories that initially start very (formally, infinitesimally) close. Already at an intuitive level, this tells us that a chaotic system, once regarded as source of messages (e.g., the trajectories, or their representation as sequences of real numbers), produces information with a finite rate. An easy way to convince you about this is to imagine that you see your trajectory with some level approximation (e.g., you do not distinguish two real numbers closer than a given threshold, whatever small, value ϵ). Due to the sensitive dependence on initial conditions it means that when recording two trajectories initially at distance much

---

[6]Formally, $I(X_{in}, X_{out}) = H(X_{in}) + H(X_{out}) - H(X_{in}, X_{out})$, i.e., the mutual information between input and output is just the sum of the entropy of the input and the output minus the joint entropy of the concatenation of input and output. If input and output are independent $H(X_{in}, X_{out}) = H(X_{in}) + H(X_{out})$ and thus $I = 0$ as expected.

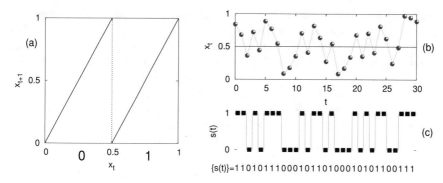

Fig. 15.3 Bernoulli shift map. **a** Graphical representation of the relation between x_t and x_{t+1} and illustration of the coding. **b** Example of a trajectory starting from a generic initial condition. **c** Symbolic sequence representing the trajectory in (**b**)

smaller than ϵ, even if at the beginning the sequences of numbers they generate are the same (as observed with resolution ϵ), soon their distance will be larger than ϵ and thus you will see two (mostly) independent sequences emerging from initially identical ones, and thus the number of possible sequences will grow exponentially. Since the growth of the distance is (exponentially) controlled by the Lyapunov exponent, intuitively we can guess that the Lyapunov exponent should have something to do with information production and thus with entropy. To make such a connection clearer we will consider a very simple illustrative example (Fig. 15.3).

One of the simplest chaotic systems is the so-called Bernoulli shift map, which is a discrete-time system which evolves a real number x with $0 \leq x \leq 1$ according to the following elementary rule. Take x_t as the value of the number at time t, multiply it by 2 if $2x_t < 1$ then $x_{t+1} = 2x_t$, while if $2x_t > 1$ then take $x_{t+1} = 2x_t - 1$ (see Fig. 15.3a for a graphical representation).[7] The good point with this map is that you can exercise on it using a pocket calculator. If you look at the evolution of x_t starting from a *generic*[8] initial x_0, as shown in Fig. 15.3b, you will not recognize any regularity and you will call the trajectory disordered or random, as typical for chaotic systems. Clearly, two initially very close initial conditions x_0 and $x_0' = x_0 + \delta$ in a step will be at distance 2δ, in two steps at distance $2^2\delta$, etc. That is their separation will grow as $2^t = \exp(t \ln 2)$ (where ln denotes the natural logarithm), meaning that the Lyapunov exponent of the system is $\lambda = \ln 2 > 0$, as it should be for a chaotic system. Now let us transform the trajectory into a sequence of (a finite number of) symbols. For instance, we can code the trajectory $x_0, x_1, \ldots, x_k \ldots$ into a sequence of symbols $s_0, s_1, \ldots, s_k, \ldots$ by choosing $s_k = 0$ if $x_k \leq 1/2$ and $s_k = 1$ if $x_k > 1/2$. With reference to the figure, the chosen partition is shown in panel (a) and the symbolic sequence associated with the trajectory in panel (b) is shown in panel (c). We are basically doing a partition of the interval of possible values into two intervals

---

[7]In a compressed mathematical form this map can be written as $x_{t+1} = 2x_t \bmod 1$ with $x_0 \in [0, 1]$.
[8]The word *generic* is not innocuous here and has a precise mathematical meaning that at least intuitively will be clear in a few lines.

and assign a symbol to each of them. With this coding the trajectory is converted in strings of 0s and 1s and, for the specific map under consideration, one can show that these strings are random as if they were generated by a fair coin tossing game, i.e., 0 and 1 appears with probability 1/2 and two consecutive outcomes are completely independent. This is quite striking as we generated the sequence with a deterministic rule (to better understand the origin of this randomness the reader is invited to read the entry Complexity of this book). Being 0 and 1 independent and equiprobable we can understand that the entropy per symbol of the messages emitted by the chaotic map will be 1 bit ($-\log_2(1/2) = 1$) which, aside from the change of the logarithm base from the natural to 2, is exactly the Lyapunov exponent of the map. This result derived for a specific example is actually general: chaotic systems are sources of information and the entropy rate (per symbol in the discrete-time systems, or per unit time in time continuous ones) is called Kolmogorov-Sinai (KS) entropy (see the entry Kolmogorov to deepen his contribution to chaos and information theory) which is linked to the positive Lyapunov exponents.[9] Importantly the KS entropy is an invariant characteristic of the chaotic system (i.e., the source), as the Lyapunov exponents are, meaning that it is an intrinsic property not affected by the change of variables. We underline, however, that the KS entropy to be properly defined and to enjoy the above properties should be computed not for a generic partition (i.e., the coding is not arbitrary). The partition of the unit interval we used for the Bernoulli shift map looks natural, and we have been lucky that it is also a choice which ensures a very important property, namely, given an infinite sequence of 0/1 symbols obtained by it we can reconstruct with arbitrary precision the initial condition. When this happen the partition is called *generating* and the entropy computed from it equals the KS entropy. When the generating partition is not known the KS entropy is defined by considering the entropy of a generic partition and taking the supremum over all possible partitions. This definition, however, is not constructive. In practice, a way to estimate the KS entropy is to consider very refined partitions. For instance, for the Bernoulli shift map one can divide the unit interval in $N = 1/\epsilon$ subintervals of size ϵ, compute the Shannon entropy $h_{Sh}(\epsilon)$, the KS entropy will then be taken as the limit $\lim_{\epsilon \to 0} h_{Sh}(\epsilon)$.

We would like to stress once again that the Shannon entropy (or for chaotic sources the KS entropy) characterizes the rate of information production of the source and not of a specific message. The Bernoulli shift map offers the possibility to illustrate this with a simple example. Consider the initial condition $x_0 = 1/3$, clearly $x_1 = 2/3$ and $x_2 = 4/3 - 1 = 1/3$, etc. In other terms with this (non-generic) initial condition the map will give rise to a periodic alternation between 1/3 and 2/3, consequently the digital sequence of 0s and 1s originating from this trajectory will be 01010101010... which is not random. You can convince yourself that this happens with all the rational numbers, thus for an infinite (but countable) number of initial conditions. Still the entropy of the source is 1 bit. As said, the Bernoulli map is equivalent to a coin

---

[9]For the Bernoulli shift map, which has only one positive Lyapunov exponent, the KS entropy is equal to the Lyapunov exponent. In general, for chaotic systems with more than one positive Lyapunov exponents, the KS entropy is not larger than the sum of the positive Lyapunov exponents.

tossing, and flipping a coin in principle the sequence of heads(1)/tails(0) of the form 01010101... is (strange but) not forbidden, yet the Shannon entropy of the fair coin tossing game is 1 bit. This shows the inadequacy of the entropy in characterizing the "complexity" of single messages, see the entry Complexity Measures for a discussion of this aspect.

Chapter 16
Irreversibility

Imagine to take a deck of playing cards and order them, for instance, group the different suits and then for each suit sort the cards from the lowest to the highest value (say, from ace to king). This job requires a certain effort and takes time to be completed. Once completed, shuffle the ordered deck by hands, in the common way. It is a familiar experience that a few quite repetitive and simple gestures of the hands are sufficient to destroy the order in the deck (see Fig. 16.1). Perhaps, to avoid unpleasant small groups of ordered cards, one needs more shuffling cycles. Ten cycles are usually enough. It is a matter of few seconds.

Our experience tells us also that it is impossible to obtain an ordered deck of cards by shuffling a disordered one. Playing cards remind us of a fact that appears in many other moments of our everyday life, i.e., that it is very difficult (and it costs some work) to obtain order and very easy to wash it out: think of your kids spending five minutes in their room just made up! Roughly speaking, this fact is the essence of irreversibility, an ubiquitous concept in fundamental research, such as in physics, but also in applied sciences, such as biology or chemistry. There are many transformations that—in nature—spontaneously occur in only one direction. Life is an evident example, though a very complex one: if we exclude fiction (a famous tale is Francis Scott Fitzgerald's "The curious case of Benjamin Button", adapted into a movie by David Fincher in 2008) it certainly goes from birth to death.[1]

Irreversibility can also be experienced in many simple physical phenomena. For instance, we never see heat flowing spontaneously from a cold body to a hot one: a fire heats a pot full of water and eventually make it boiling, but it is ridiculous to expect that a pot full of boiling water gets colder making warmer the fire. Similarly, a gas contained in a balloon, when the balloon is pierced, gets out and spreads everywhere

---

[1]Life is always surprising, for instance, *Turritopsis dohrnii*, also known as the immortal jellyfish, is one of the very few exceptions to the rule as it is able to revert completely to a sexually immature stage after having reached sexual maturity.

© Springer Nature Switzerland AG 2021   105
M. Cencini et al., *A Random Walk in Physics*,
https://doi.org/10.1007/978-3-030-72531-0_16

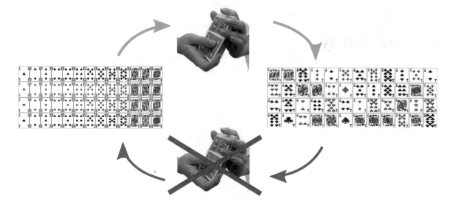

Fig. 16.1 Cards can be shuffled from ordered to disordered. Our experience suggests that reversed shuffling is impossible

in the room, the opposite (the gas from the room inflating the balloon) is certainly unexpected.

In summary, irreversibility is a familiar and (apparently) intuitive phenomenon. There is, however, a fascinating problem lying behind it: a deep, or hidden, paradox, which is usually overlooked and, conversely, makes the scientists insomniac. In the context of theoretical physics, the problem is that irreversibility implies a temporal asymmetry which is at odds with the fundamental laws being time symmetric. Let us see why.

In the popularization of science as well as in technical papers, irreversibility is often represented through a simple image: an arrow. In fact "the arrow of time" is a widespread expression in physics. It effectively conveys the meaning of phenomena that occur in only one direction, as in a one-way street. Irreversible phenomena, when seen happening in the opposite direction, are immediately recognized as false, impossible, and unreal. A video showing the surface of a table where a large spot of milk shrinks and then jumps into an inclined cup is evidently played in the wrong direction, from the future to the past. Irreversibility is an asymmetry in the spontaneous evolution of many natural things. In order to make this idea of a "time arrow" formal and quantitative, physicists use a famous physical concept: entropy. This is described in the entry Entropy, where we explain how it quantifies the disorder of a physical system made of a very large number of atoms. Entropy is one of the proper quantities to quantify irreversibility. The common experience of events that occur in a well-defined time direction is translated, in physics, into the Second Principle of Thermodynamics, which states that entropy cannot decrease spontaneously; in fact, excluding very slow transformations (where it remains constant), entropy of an isolated system can only grow. Spontaneous processes such as the expansion of a gas (e.g., from the balloon to the room) or the conduction of heat from cold to hot bodies are typical examples, in a lesson of thermodynamics, where entropy grows.

So far, we have understood that irreversibility is a strong and pervasive example of asymmetry in time. Where is the paradox? The fact is that physicists—since more than a century—know for sure that all things in the universe are made of atoms (see the entry Atoms). Atoms move according to well-defined and understood laws: the laws of mechanics. It is not important here if we consider classical mechanics (which is a good approximations at the scale of our everyday life, i.e., if the objects are not too small) or we stick to quantum mechanics, which is more generally true but often very complicated. Being them classical, relativistic, and/or quantistic, all mechanical laws do not distinguish the past from the future, i.e., any evolution of atoms that obeys these laws can be "time-reversed" (this is like reading a story from the end to the beginning) to produce a different evolution that still obeys the same laws. As written by Brian Greene in his book *The Fabric of the Cosmos: Space, Time and the Structure of Reality*, *"no one has ever discovered any fundamental law which might be called the Law of the Spilled Milk or the Law of the Splattered Egg"*.[2]

In summary, there is no arrow of time in the laws of mechanics that is in the movement of atoms of a macroscopic body. This implies two different consequences. First, it is not possible to distinguish the correct direction of a movie displaying moving atoms (basically there is no a priori correct direction). Second—and most importantly—the above unphysical example of milk that *unspills* from the table to a cup is perfectly compatible with the laws of mechanics. If we could magically reverse the velocities of all the molecules of milk spread on the table, those molecules would trace back their path to their original positions, without violating any mechanical law. Surprisingly, for every phenomenon observed in nature there exists its "time-reversed twin" which is perfectly allowed by mechanics: simply, we never observe it! If the reader reflects some minutes about this point, then she will be really puzzled.

For physicists, this problem represents an apparent profound incompatibility between the microscopic world ruled by symmetric mechanical laws and the macroscopic world (that of milk, gases, or fire) which, conversely, appears to be ruled by asymmetric laws such as the Second Principle of Thermodynamics. Somehow, the Second Principle contradicts the laws of mechanics, as it states that certain evolutions of atoms—dictated by mechanics—are instead forbidden. Equivalently, spontaneously decreasing entropy is perfectly compatible with the laws of mechanics.

This paradox was solved by Ludwig Boltzmann (see the entry Boltzmann), although—interestingly and remarkably—even today there are scientists who are not yet convinced by Boltzmann solution or consider it counter-intuitive. A famous case is that of Ilya Prigogine (1917–2003), Nobel Prize for chemistry in 1977: he often claimed that irreversibility cannot be true unless it is true also at the level of

---

[2]The word "fundamental" here is rather controversial and should be replaced by "mechanical" or "microscopic". A part of the scientific community considers "fundamental" only the laws of mechanics, since they are obeyed by the elementary units of matter, and thinks that every other natural phenomenon can be understood just in terms of these laws. Another part of the scientific community, on the contrary, thinks that there are "fundamental" laws of physics which cannot be reduced to the laws of mechanics, but which emerge at different levels of description, for instance, when the number of elementary constituents becomes very large. The Second Principle of Thermodynamics, which in this case could be called the Law of Spilled Milk, is exactly one of these.

atoms. The fact is that the solution proposed by Boltzmann requires the introduction of probability, or statistics, which may seem to contradict the idea of determinism, intrinsic in mechanical laws. Under the hypothesis that the state of the system at a given time is perfectly known, classical mechanics establishes an exactly predictable evolution. Mechanics was conceived to understand the motion of planets, it is able to design complicate machineries in watches measuring time and to send ships into outer space, making two of them encounter in the right spot of the infinite sky, etc. So, why (and how) should we introduce probability into it?

Probability, in fact, is essential in describing the evolution of systems made of many things, such as atoms or playing cards. Let us go back to our initial example. The act of shuffling cards is both neutral and not neutral, from the point of view of order. It is neutral because each movement of the hands can bring any order of cards, without preference: it is perfectly possible that it transforms a disordered configuration into an ordered one. Figure 16.1—in a sense—is false: there is nothing that prevents the passage from disorder to order through the (red crossed) "reverse shuffling". However, there is a difference between order and disorder: while a disordered configuration is *not unique*, there are a huge number of disordered card arrangements, conversely the ordered ones are very few. Shuffling cards and getting an ordered configuration would be an incredibly lucky event. So the figure where the reverse shuffling is red-crossed is substantially true, it is practically impossible to observe it.

The example of the gas is also quite simple to be understood: it suffices that we recall the microscopic nature of a gas substance that is a (very large) bunch of tiny "flying balls", the molecules. These crazy balls move just like in a huge billiard, colliding among themselves and with the walls of the room, never stopping: it is not difficult to believe that the probability of them going all together—through a tiny hole—into a small balloon is essentially zero. Transferring heat from cold to hot is again a matter of ridiculously small probabilities: we should wait for the few fast molecules in the cold water to go to touch the few slow molecules in the fire and hope that this incredibly lucky event occurs much more frequently than the opposite, transferring energy from the cold to the hot body.

Let us make some easy numerical estimate, to make these arguments more convincing. In a poker deck, there are 52 cards. It seems a small number; however, we find very intuitive the fact that going from disorder to order is terribly unlikely. Our intuition is explained through the simple comparison between the numbers of ordered and disordered configurations. What about the number of "ordered" arrangements of cards? We must decide what an ordered arrangement is. If we consider strong orders, like in the example at the beginning (cards grouped in suits and in order of ascending value), we have only 24 possible configurations (because $24 = 4!$ are the possible arrangements of the 4 suits[3]). We can be more tolerant and call "order" other kinds

---

[3]We recall that the notation $n!$ in mathematics represents the number $n \cdot (n - 1) \cdot (n - 2) \cdot ... \cdot 1$ and is called factorial. The ! symbol expresses *wonder* and is very convenient in this case; in fact $n!$, when n increases, becomes rapidly astronomically large, as clear from the examples given in this chapter.

of strange, unnatural, configurations. For instance, we can call ordered all configurations where the cards are just grouped in suits, without each suit being internally ordered according to values. Such configurations are 13!4! which is a number with 11 digits, which seems to be large... But what about the number of disordered card arrangements? Consider that the number of all possible arrangements of poker cards is 52!, i.e., a number with 67 digits! First, this means that subtracting from such a huge number the number of ordered configurations does not change the number itself. And therefore the ratio between the number of ordered configurations and the disordered ones is ridiculously small, is basically 0.000...., and after roughly 50 more zeros a final ...0001. If associated to a probability, this is not distinguishable from zero.

Now, try to imagine what happens with a gas of molecules: in Fig. 16.2 we sketch an example with only five molecules.

In the reality, the number of this tiny balls in a balloon is immense, for instance, in a cube centimeter of gas the number of molecules is 10 followed by 19 digits. And each ball is not just a card which can take 52 values, but is a molecule which can be in a very large number of different places (more than the few tens sketched in the figure above). When the balloon is pierced, the laws of mechanics start "shuffling" these balls from the original positions, making them explore the rest of the room. The configuration of molecules in the balloon, of course, is not unique, many similar arrangements of molecules are equivalent to the state labeled "in the balloon": let us call the number of these configurations W_b (in theoretical physics this number can in fact be counted, in the example depicted in the figure above we can do it easily). In principle, we can also count the number of configurations equivalent to the state labeled "in the room", and call it W_r. The fundamental fact is that W_r is incredibly much larger than W_b. The ratio between the number of "room" configurations W_r and the number of "balloon" configurations W_b is larger than the ratio between the volume of the universe and the volume of a single atom! The probability to observe by chance the gas in the balloon is the reciprocal of this number and is immensely small, indistinguishable from 0. Of course, classical mechanics is deterministic and from each configuration in the balloon the gas evolves—after some time—into a perfectly determined configuration: all W_b balloon configurations are transformed into W_b final configurations. Now you are asked to bet whether the W_b final configurations belong to the original W_b balloon ones, or to the W_r room ones. Which one would be your bet? Everyone would bet that they belong to the room ones, because they are vastly more than the baloon ones, so that there is an extraordinary larger probability that the bet will be won.

As discussed in the entry Entropy, Boltzmann understood that thermodynamic entropy, which is related to the flow of heat, basically counts (even if in a "logarithmic" way) the number of atomic configurations of the gas, or of any other substance. Therefore, the above quite convincing argument about the extremely high probability to go from a state realized in W_b possibilities and a state realized in W_r possibilities, with W_r/W_b an enormous number, is also a quite convincing argument about the necessity of increasing entropy, namely, for the second principle of thermodynamics. Somehow we get to the conclusion that the second principle is both true and

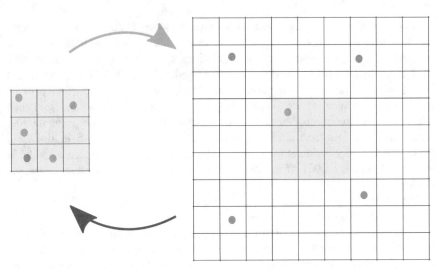

Fig. 16.2 The expansion of a gas made of 5 molecules, from the closed balloon (left, small volume) to the room (right, a volume 9 times larger than the balloon): the open balloon is still represented in the room as the yellow area. There are $W_b = 9 \cdot 8 \cdot 7 \cdot 6 \cdot 5 = 15120$ different ways of arranging the molecules in the balloon. There are however $W_r = 81 \cdot 80 \cdot 79 \cdot 78 \cdot 77 = 3'074'591'520$ (more than three billions!) ways of arranging the molecules in the room. Finding, just by chance, a configuration of molecules in the balloon has a probability W_b/W_r which is smaller than 0.001%. When the number of molecules is 10^{20} (as in reality) and the real volumes are taken into consideration, this probability becomes indistinguishable from zero

false: the laws of mechanics allow for the possibility of violating this principle, but the probability to see such a violation is indistinguishable from zero.

The growth of entropy for systems made of many atoms that follow deterministic mechanical laws is a fact with several subtle and counter-intuitive consequences, and that is the reason why it is still debated and hardly digested by students. For instance, one could object the following: if I take a configuration of all positions and velocities of the molecules in the balloon and magically change the sign of all the velocities, I will see the molecules go "back in time". Does it mean that I should see entropy decreasing? Absolutely not, but it is not obvious to make it clear through simple examples. A basic explanation is the following. In order to observe a decrease of entropy we should observe a reduction of the volume occupied by the gas. In other terms, we should see the gas molecules concentrating in a region even smaller than the balloon. But we can be pretty sure that *any* initial configuration of the molecules' positions, whatever the velocities are, will bring the gas out of the balloon, spreading around the room, just because the number of possibilities of doing so is vastly bigger than those of focusing in a small region. In the very first instants—for a ridiculously small amount of time after the rupture of the balloon—the gas could do something unusual, counter-intuitive, even against the "arrow of time". Such very brief unusual behavior may occur with any initial configuration but is substantially not observable, its duration is as small as the probability discussed above, we are speaking of numbers of the kind 0.0000.... followed by a very large number of zeros and a ..001 at the end. A gas in a balloon that has been punctured an instant ago, from the point of view of entropy, is like a ball on the top of a very pointed, sharp, mountain: whatever direction it takes, it will go toward the valley (here the metaphor needs a change of sign, going down for the ball is equivalent to going up for the entropy). If you have a powerful magnifying glass you could realize that the top of the mountain has small asperities and therefore the first instants of the "going down" could be, in fact, temporary "going up" but for extremely small amounts of time. Seen from a far point of view, anyway, the ball goes down, with no doubts.

Some devil's advocate could make a nasty objection. A few pages above we have recalled that the phenomenon of milk that unspills from the table to the cup is compatible with mechanics and, in fact, can be realized by just reversing the velocities of the atoms of spread milk. Are we contradicting ourselves? Again, no, but it is not easy to convince ourselves of it. We are not saying that it is not possible to reduce entropy by inverting the velocity of all molecules, but that it can happen only with a very small probability. To see the molecules in the balloon go in a much smaller region, we need to pick up a configuration of the molecules that has evolved (spreading) from that small region. Only in that case, inverting the velocities would bring back the molecules in the small region, thus reducing entropy. But the number of such configurations, let us call it W_s ("s" = "small"), is incredibly smaller than the number of all possible configurations of molecules in the balloon, W_b: the reasoning is exactly the same as in the case of spreading from the balloon to the room. In that case we discovered that there is an immense unbalance between the W_b and W_r. For the same reason, there is an immense unbalance between W_s and W_b and this immediately dissolves the present objection.

A further consequence concerns the whole universe and, therefore, cosmology, whose subject is the origin and evolution of the universe. Remember that irreversibility (i.e., the second principle of thermodynamics) affects spontaneous processes or—more precisely stated—isolated systems, i.e., any system where there are no external actors that can spend their energy and force the system to do *strange* things. In fact, there are many simple cases of *non-isolated* systems where strange things happen: think of a fridge, it is a machine that makes heat flowing from cold to hot. It creates order (cold) from disorder (hot) but it is not violating the second principle because it spends a lot of energy, it is certainly not a spontaneous process. If we consider a bigger system made of all things involved in fridge's operations (the fridge itself, the surrounding environment, the power grid, and its original source of energy, such as a waterfall, etc.) we get a more complete account of total entropy changes, and we can only see entropy growing, the total system is becoming more disordered. This is the reason why the appearance of life does not contradict the second principle: living beings are very complicate and non-random systems, life is certainly more ordered than the primordial chaos of a planet just emerged from dust cooling in outer space; however, a planet is not an isolated system but constantly receives and dissipates energy from its sun. Life can appear on a planet, making entropy decrease in some tiny region of space and time. But the total entropy of the whole universe, the system that includes everything and is isolated by definition, can only increase. This observation, according to many scientists, implies that the universe—at its origin, for instance the Big Bang—was in a state with an entropy much lower than today. Perhaps Big Bang was like a gas compressed in a very small balloon which has been pierced just a moment before.

Surely the most important attempt to conciliate mechanics and thermodynamics is due to Boltzmann (see related entry) who in his path toward the understanding of irreversibility faced the two following extraordinary problems:

• to obtain, within Newtonian mechanics, the Maxwell-Boltzmann distribution for the velocity of the molecules of a gas of mass m in thermodynamical equilibrium at temperature T:

$$p_{MB}(\mathbf{v}) = const.\, e^{-\frac{m\mathbf{v}^2}{k_B T}} \, ,$$

where k_B is Boltzmann's constant;

• to show, within classical mechanics, that starting with any probability distribution of the particle velocity $p(\mathbf{v}, 0)$, after a certain time t, $p(\mathbf{v}, t)$ becomes close to the Maxwell-Boltzmann distribution. For a diluted gas, Boltzmann was able to derive, with a suitable hypothesis (molecular chaos), the evolution equation (now called Boltzmann's equation) for $p(\mathbf{v}, t)$. From Boltzmann's equation, it follows that there exists a stationary[4] solution $p_s(\mathbf{v})$ given by the Maxwell-Boltzmann distribution $p_{MB}(\mathbf{v})$, and the evolution takes every $p(\mathbf{v}, 0)$ initially different from $p_{MB}(\mathbf{v})$ toward it.

From his equation Boltzmann was able to show the celebrated H theorem: the quantity

$$H(t) = \int p(\mathbf{v}, t) \ln p(\mathbf{v}, t) d\mathbf{v}$$

cannot increase during the time evolution and reaches its minimum for $p(\mathbf{v}) = p_{MB}(\mathbf{v})$. It is possible to realize that in a dilute gas the number of possible molecule configurations is strictly related to the number of possible configurations of the velocity of a single molecule (the underlying reason is because the molecules are statistically independent). Therefore, the quantity $H(t)$, apart from a minus sign, counts the number of all possible configurations of the gas and, therefore, we have the relation $H = -S/k_B$, being S the entropy. Boltzmann thought that his H theorem was a genuine mathematical proof of the second law of the thermodynamics.

Unfortunately, the original solution proposed by Boltzmann to the problem of irreversibility clashes with two consequences of the Newton's dynamics, namely, the recurrence paradox (due to E. Zermelo) and the reversibility paradox (due to J. Loschmidt). Zermelo noticed that the H theorem disagrees with Poincaré's recurrence theorem, whereby in any mechanical system whose motion takes place in a bounded region of phase space, at some (possibly large but) finite time the system will return arbitrarily close to the initial condition. Since the distribution function $p(\mathbf{v}, t)$ (and thus $H(t)$) depends on the velocity of the molecules, when the system

---

[4]Here stationary means that while the system is evolving in time, the probability distribution $p_s(v)$ remains the same.

returns close to its initial state, H must reach a value close to $H(0)$. Consequently, $H(t)$ cannot be a monotone function of time. As for the reversibility paradox, if $H(t)$ decreases from time 0 to time t, and at t the velocities are reversed then, because of the time-symmetric nature of the equations of motion, the system will go back after t, and H should increase.

The best way to address the above objections is to observe that the H theorem cannot be considered a theorem of classical mechanics, and hence it does not hold for any solution of the equations of motion. According to Boltzmann, *"the second law [of thermodynamics] cannot be deduced mathematically from the equations of dynamics alone (...). What I proved is that it is highly likely that $H(t)$ will approach its minimum; if it is bigger, it can either increase or decrease, but the probability that it will decrease is [much] higher"*.

Boltzmann managed to give a preliminary answer to the paradoxes uncovered by Zermelo and Loschmidt (and later his arguments would be proved to be essentially correct). As for the recurrence paradox, he pointed out that Poincaré's return time is for macroscopic systems extremely long, and *de facto* not observable. For example, before all the particles in one cubic centimeter of gas at normal pressure and temperature return (close) to their original configuration, say with a precision of 10^{-9} m on the position and 1 m/s on the velocity, we would have to wait something like $10^{10^{19}}$ years, incommensurably much larger than the universe's age (which is estimated "only" in the order of 10^{10} years). Today Boltzmann's argument is subsumed by Kac's lemma, discussed in the entry Big Data.

Now we can say that Boltzmann's intuitions are correct, perfectly confirmed by numerical simulations as well by rigorous mathematical results. We know that the H theorem is correct, at least for a certain time and in a suitable limit, for instance, in the so-called Grad-Boltzmann limit, i.e., a system of elastic spheres of diameter σ, when the number of particles per unit volume, N, tends to infinity while σ tends to 0, in such a way that $N\sigma^2$ is constant.

Chapter 17
Kolmogorov

Andrei Nikolaevich Kolmogorov (1903–1987, ANK in the following) was perhaps the foremost contemporary Russian mathematician and among the greatest scientists of the twentieth century. He represents, with Hilbert and Poincaré, a counterexample to Bourbaki's observation that *"Even among those [mathematicians] who have the widest training, there are none who do not feel lost in certain regions of the immense world of mathematics"*. His mathematical horizon, indeed, encompassed almost all areas of mathematics. As a far-from-exhaustive list of mathematical subfields in which he named some seminal contribution, we can mention the theory of trigonometric series, measure and set theory, constructive logic, topology, probability theory and the theory of random processes, information theory, ergodic theory, dynamical systems, celestial mechanics, differential equations, and the theory of algorithms. ANK did not enclose himself in the tower of pure mathematics and gifted science with very influential works in mathematics applied to biology, genetics, geology, and ballistics. Moreover, he demonstrated great physical insights in many problems and, in particular, he laid the foundation of modern approaches to turbulence theory (see the entry Turbulence). As a recognized hallmark of his scientific attitude, we can say that in all his contributions, even the shortest ones, he did not focus on a single question but did always take wide breath providing fundamental insights and deep links among different disciplines, often pioneering new fields of investigations. In this brief sketch of his scientific and personal life, it will be impossible to touch all the themes and aspects of his research, so we will mainly focus on those which highlight the extent of his interest in some of the subjects discussed in other entries of this book and provide some glimpse of his personality as a man and as a scientist.

ANK was also a great teacher, he supervised more than 60 Ph.D. students many of which become among the most celebrated scientists of the twentieth century. His teaching attitude is well summarized by the words of one of his most famous students, V. I. Arnol'd, *"K. never explained anything, just posed problems, [...] He gave the student complete independence and never forced one to do anything, always waiting to hear from the student something remarkable. He stood out from the other*

© Springer Nature Switzerland AG 2021
M. Cencini et al., *A Random Walk in Physics*,
https://doi.org/10.1007/978-3-030-72531-0_17

professors I met by his complete respect for the personality of the student". In 1963, he promoted the foundation of the mathematical boarding school N. 18 at Moscow University, known as the *Kolmogorov school*, for talented high school students that, even today, serves as a model for specialized high schools at prominent universities. To give an idea of his engagement in education we mention that, for several years, he was giving 26 hours of lessons a week providing also written notes. Moreover, he was also teaching students on music, art, and literature. Students from his school were among the most successful in international mathematical olympiads.

ANK was born on April 25, 1903 in Tabov in Russia. His mother, Mariya Yakovlevna Kolmogorova, died in child-birth, his natural father Nikolai Kataev abandoned him not caring of his upbringing. He was grown by the sister of his mother, Vera Yakovlevna Kolmogorova, who, in his words, *"raised him in the sense of responsibility, independence of opinion, intolerance towards idleness and poorly performed tasks, and the desire to understand and not just to memorize"*. The mother family was of noble origin, fact which caused some difficulties after the Russian revolution. He spent the childhood in the family estate in Tunoshna. With his aunt he moved to Moscow in 1910 where he finished the high school in 1920. After the school, he worked for some time as a railway conductor.

In autumn 1920, he became student at Moscow University. Those were hard times in the new URSS, lecture rooms at university were unheated. When ANK discovered that second year students were receiving 16 Kg of baked bread and 1 Kg of fat a month more than first year ones, he checked the minimum requirements and quickly passed all the exams needed to move to the second year. At the beginning of university, he was very much interested in Russian history, field in which he wrote a treatise on the fifteenth-century history of the Russian city Novgorod. Then he completely turned to mathematics under the influence of Prof. Stephanov and the supervision of Prof. Lusin. In 1922, he obtained the first important mathematical result demonstrating, by construction, the existence of an integrable function whose Fourier series diverges almost everywhere. This result gave him international recognition, in spite of his young age. He graduated in 1925. Since 1924, ANK started to be interested in probability theory, moving his first steps in this field with A. Y. Khinchin. From 1924 to 1928, he established the necessary and sufficient conditions for the validity of the strong law of large numbers and proved the law of iterated logarithms for the sum of independent random variables. ANK finished the Ph.D. in 1928, at age 25, with 18 mathematical papers published.

In 1929, ANK started his lifelong friendship with P. S. Aleksandrov, another important mathematician, who had an enormous impact in his personal and scientific life, in ANK words (in 1982): *"for me these 53 years of close and indissoluble friendship were the reason why all my life was on the whole full of happiness, and the basis of that happiness was the unceasing thoughtfulness on the part of Aleksandrov"*. Their friendship started during a long summer vacation during which the two spent together sailing and camping along the Volga using a boat and tent provided by the *Society for Proletarian Tourism and Excursions*. They spent 3 weeks in rivers, lakes, and hiking on mountains (climbing up to the 4100 m of Alagez mountain). Of course, while in vacation, they were also doing mathematics: ANK working on integration theory and analytic descriptions of Markov processes in continuous time, and Aleksandrov on his book on topology written with Hopf. Close to the end of their vacation they decided to share a house.

During the years 1930–1931, Kolmogorov and Aleksandrov were mainly abroad. In 1930, they were both in Göttingen, where ANK had the possibility to discuss with H. Weyl on intuitionistic logic, with R. Courant on limit theorems, and with E. Landau in function theory. In summer, they were invited by M. Frechet, with whom ANK discussed about probability theory, Kolmogorov's main research interest in those years. After they moved to Paris, where ANK had contacts with E. Borel and P. Lévy. In 1931, before going back to Moscow, ANK went back to Göttingen and Aleksandrov in the USA. During 1931, ANK published *Analytical methods in probability theory*, which basically gave birth to the modern theory of Markov processes. In the words of Gnedenko: *"In the history of probability theory it is difficult to find other works that changed the established points of view and basic trends in research work in such a decisive way. In fact, this work could be considered as the beginning of a new stage in the development of the whole theory"*. The same year he became professor at Moscow State University, and from 1937 held the chair of probability theory. In 1933, he published in German the *Foundations of the Theory of Probability*, which can be considered the most important contribute of Kolmogorov in the first part of his career. In this book, he laid the foundation of the modern axiomatic foundations of probability theory leaving behind any dispute on the interpretation of probability, as between the *frequentist* and *subjectivist* ones. The entry Probability covers these aspects.

In 1935, Kolmogorov and Aleksandrov established in a "dacha" at Komarovka, a large house in the countryside with a large library and room for several guests. Since then, *"As a rule, of the seven days a week, four were spent in Komarovka, one of which was devoted entirely to physical recreation—skiing, rowing, long excursions on foot (these long walks covered on average about 30 km, rising to 50; on sunny March days we went out on skis wearing nothing but shorts, for as much as four hours on a stretch. On the other days, morning exercise was compulsory, supplemented in the winter by a 10 km ski run [...] Especially we did love swimming in the river just as it began to melt [...] I swam only short distances in icy water but Aleksandrov swam much further. It was I however who skied naked for considerably longer distances"*. These words of ANK very well illustrate the importance he gave to physical exercise besides mathematics. The house of Komarovka became a meeting place for many mathematicians and students, and played an important role in the Russian school of mathematics.

The interest of Kolmogorov for probability theory was not limited to formal and technical aspects, he really changed the point of view on the subject and he had the main responsibility for laying the foundation of the modern theory of stochastic processes, which led him to apply these ideas to physical and biological problems. His contributions in these fields were not mere mathematical answers to specific problems but led to entire new fields of investigations or revolutionary understanding of the problem, as in the case of turbulence.

On the side of mathematical biology, in 1936, Kolmogorov generalized the Lotka-Volterra model of predator-prey systems and wrote an influential paper in Italian, in honor of Vito Volterra (see the entry Volterra). One year later, together with Petrovsky and Piscounov, he introduced (simultaneously and independently of R. A. Fisher in England) a model for describing the spreading of an advantageous gene. This work (together with Fisher's one) besides having an important impact in genetics and ecology laid the foundation of modern reaction-diffusion processes. Generalizations of this model are currently used in applications encompassing population biology and genetics, epidemic spreading, complex chemical reactions, pattern formation, and combustion. It must be emphasized that ANK "incursions" in biology were not done as an amateur. He was always going deep. For instance, Kendal remembers that Kolmogorov, who was invited in a meeting at Oberwolfach (in 1967) on the analytical theory of branching processes, *"At first he explained that he only wanted to be a listener, but at the end of several highly mathematical talks he looked rather uncomfortable, and eventually told us that he would after all give a talk that would perhaps remind people of the biological background to the subject"*. Moreover, his strong opinions on the subject led ANK to an open contrast with Lysenko.[1] In 1939, N. I. Ermolaeva, a protégé of Lysenko, published the results of an experiment partly repeating Mendel's ones. Analyzing the data, she concluded that Mendel's principal law of inheritance was false. The year later, ANK reached a different conclusion using the same data and exploiting the non-parametric statistical test—now known as Kolmogorov-Smirnov test (1933)—to quantify the distance between an empirical probability distribution and a reference one. He wrote the paper *On a new confirmation of Mendel's laws*, which put him in a open confrontation with Lysenko who (denying the existence of genes) replied in the same year with a paper titled *In response to an article by A. N. Kolmogorov*.

At the end of 30s, ANK become interested in the theory of turbulence, namely, to the highly irregular motion of strongly stirred fluids. In 1941, he published two short papers on turbulence introducing the concept of local structure of turbulence, scaling, and small-scale universality, and an exact result (actually one of the really few exact results that can be obtained from the Navier-Stokes equations) that radically changed the understanding of the subject. His theoretical description of fully

---

[1]Trofim D. Lysenko (1989–1978) was biologist and agronomist. His name was initially associated to a technique to obtain winter crops of cereals. He did not limit his interest to practical aspects and entered the scientific debate on the principles of genetics and hereditary transmission (such as Mendel's laws) with very anti-scientific views by re-proposing old and wrong ideas of Lamark. Unfortunately, he had the support of the communist party and Stalin himself, this gave him much power inside URSS academia, and he basically destroyed genetics as a science in Russia. For instance, the famous geneticists N. Vavilov was sentenced of death for sabotage and, even if the death sentence was suspended, he died in jail in 1943. A similar fate destroyed the career of other scholars, such as N. Tulajkov e G. Karpecenko. Lysenko's theories were calamitous for the agriculture of the Soviet Union and led to a strong reduction of cereal yields. Despite the damages he provoked, he held for a long time much power in Soviet Union academics with consequences at an international level. For instance, in the 50s, his theories were embraced by practically all directives of the communist parties at international level, which caused the exit of numerous intellectuals and scientists, who were against him.

developed turbulence was very general and with long-lasting influence in other fields. For instance, the concepts of scale invariance and universality are at the basis of the Renormalization Group, developed only in the 70s, which was fundamental both to the understanding of phase transitions and to the development of high-energy physics. In answer to an observation of Lev D. Landau (possibly the greatest Russian physicist of the twentieth century), in 1962 he went back to the problem of turbulence developing new ideas on the role of fluctuations and intermittency in turbulent flows that, though now partially outdated from a technical point of view, are still the starting point of the (not yet complete) building of the theory of turbulence. In the entry Turbulence, the reader will find more details on ANK's contribution and the modern developments. It is interesting to mention an anecdote from Ya. G. Sinai, one of his most famous students, which shows how Kolmogorov's approach to the problem was quite unexpected for a mathematician of his caliber and illustrates his uncommon way of thinking and working: *"When Kolmogorov was about 80 years old, I asked him about his discovery of the scaling law. He gave me an astonishing answer, saying that he had studied the results of experimental measurements for about a half year"*. As a further demonstration of his versatility, we can mention that at the beginning of the 70s took part in an oceanographic campaign, led by his student A. Monin, who wrote *"Andrei Nikolaevich was responsible for the geophysical oceanic investigations. [...] Some of the devices were new and did not function properly. Kolmogorov spared no effort and time in checking the measurement accuracy and the calibration of the devices as well as in revealing the interferences that distorted the readings"*. ANK himself formalized his approach when interviewed by the documentary-film maker A. N. Marutyan, saying *"The mathematicians always want that their mathematics should be pure, that is, strict and provable, wherever possible. However, the most interesting and realistic problems could not usually be solved in that manner. Therefore, it is very important that the mathematician should be able to find the approximative (not necessarily strict but effective) ways of solving such problems"*. His attitude and words clearly teach us that there is not a fundamental and an applied science, but only science and its applications.

Another research field inextricably linked to ANK is the theory of chaotic systems and ergodic theory. We can distinguish two main topics: one related to Hamiltonian systems and their application to celestial mechanics and the other related to the characterization of chaotic systems with ideas borrowed from information theory, showing his (uncommon) ability to find deep links between seemingly unrelated fields.

In 1953–1954, Kolmogorov established a first result, later refined by two of his students (namely, J. Mosers in 1962 and V. I. Arnold in 1963) about quasi-integrable Hamiltonian systems, a classical mechanics problem already identified, at the end of the nineteenth century, by J. H. Poincaré (see the related entry) while studying the motion of planets around the Sun. Very shortly, with reference to celestial mechanics (but the result applies to general Hamiltonian systems), the two-body gravitational problem (e.g., Sun-Earth) is integrable, meaning that the possible evolution of the system is either periodic or quasi-periodic (for bounded orbits)—in mathematical literature these are called *tori* as in phase space they correspond to doughnut-like manifolds. Integrability is tightly connected with the presence of conservation laws (integrals of motion) besides energy, which is strictly preserved. The presence of a third body, say Jupiter, provides a (small) perturbation—the so-called three-body problem. The presence of such perturbation raises two fundamental questions: do integrals of motion, besides energy, exist in the perturbed system? Will the trajectory of the perturbed system be "close" to those of the integrable one?

Clearly, a positive answer to the first question would imply that the perturbed system is integrable making, somehow, not necessary to answer the second question. However, Poincaré (see the related entry for details) proved that integrals of motion other than energy do not survive to generic perturbations, by a mechanism now known as the *small divisor problem*. As a consequence, as recognized already by Poincaré, chaotic orbits will appear (i.e., non-periodic and irregular trajectories from which an arbitrarily close trajectory separates exponentially in time, see entry Chaos). Before Kolmogorov approached this problem, the mathematical community was convinced that the negative answer to the first question was implying a negation of the second. Kolmogorov with a very ingenious approach, which beautifully exploits the different cardinality of irrational over rational numbers, showed that the issue is much subtler than expected. Avoiding the technicalities, the basic idea of the KAM theorem is as follows. In the unperturbed system, all orbits are periodic (or quasi-periodic)—the tori—and, thus, are characterized by specific frequencies. He showed that when the system is perturbed, if the perturbation is small, the tori with rational frequencies—called resonant tori—are destroyed while those with sufficiently irrational frequencies survive and remain close to the unperturbed ones. Since rationals are dense in the reals, they are infinitely many which implies the non-existence of global invariants of motion (i.e., Poincaré result). From set theory, however, we know that rationals have the cardinality of integers while irrational that of the continuum and, cleverly exploiting this fact, ANK proved that for small perturbations the measure of surviving tori is 1 and that they are very "close" to the unperturbed ones. This apparently technical result implies, fortunately for us, the regularity of our Solar System. Increasing the intensity of perturbation will progressively destroy non-resonant tori, until the last surviving one (i.e., that corresponding to the most irrational frequencies) get also destroyed, and chaotic orbits invade all the phase space. KAM theorem has profound implications in the way chaos appears in Hamiltonian systems, and deeply influenced the development of dynamical systems theory with applications (still ongoing) in celestial mechanics and plasma physics, where KAM tori are crucial for the problem of confinement.

The contributions of ANK to dynamical systems do not end with KAM theory, he gave a long-lasting contribution to the subject introducing, with his former student Y. A. Sinai, what is now known as the Kolmogorov-Sinai entropy of a dynamical systems. This concept was a part of his general project on information theory. ANK was, indeed, among the few mathematicians to realize from the outset the importance, not only in application, of Shannon (1948) information theory. Indeed, in a late paper in 1983, he wrote *"Information theory must precede probability theory, and not be based on it. By the very essence of this discipline, the foundations of information theory have a finite combinatorial character"*, something that, perhaps, only the founder of modern probability theory could have the right to say. In 1955, together with I. M. Gel'fand and A. M. Yaglom, he extended Shannon's ideas from discrete-time processes with discrete or real alphabets to general alphabets in both discrete and continuous time, in particular, considering time-continuous Gaussian processes. Another key contribution in this context was a generalization of the so-called rate distortion theory of Shannon. Consider a (continuous or discrete) source (which we can assume as random) emitting messages that we want to communicate. For communication purposes, the emitted messages need to be codified and transmitted. Rate distortion theory accounts for the discrepancies between the emitted (input) message and the received (output) message, due to codification and transmission. Knowing the distortion rate of information in this process is crucial to fix the transmission rate of the message through the channel so as to minimize the loss of information. ANK generalized this idea giving birth to what is now known as ϵ-entropy which turned out to be a very powerful tool to characterize the complexity of a signal at varying the scale (ϵ) of the magnifying glass through which the signal is looked at. In particular, for $\epsilon \rightarrow 0$ (infinite magnification, i.e., in terms of rate distortion theory the request of zero error between input and output signals) the ϵ-entropy corresponds to the usual (Shannon) entropy of a source.

In 1957, as reported by Sinai in his entry in Scholarpedia on the Kolmogorov-Sinai entropy, ANK gave a seminar in dynamical systems where he proposed the idea that entropy could be useful for deterministic dynamical and developed the concept for some simple system. In 1959, Sinai further developed the notion of entropy in the context of dynamical system. The key intuition of Kolmogorov was to understand that chaotic dynamical systems, when seen as sources emitting signals have a positive (but finite) entropy, which (and this is the remarkable result) is an intrinsic quantity independent of any smooth change of variables and provides a measure of the quantity of information produced by a chaotic system. In the words of Sinai: *"Following the main point of view at that time, the author [i.e. Sinai] tried to prove that the entropy is zero because the automorphism of the torus is certainly a purely 'deterministic' dynamical system. The proof didn't go through. Kolmogorov was the first to suggest that the entropy must be positive"*. Later the Kolmogorov-Sinai entropy was linked to the Lyapunov exponents (that characterize the exponential divergence of initially infinitesimally close trajectories, see the entry Chaos).

Kolmogorov's contributions to information theory were surely triggered by his uninterrupted interest on randomness and complexity produced by random or deterministic systems, which culminated in the idea of *algorithmic complexity* in the 1960s. In 1983, ANK wrote: *"The concepts of information theory as applied to infinite sequences give rise to very interesting investigations, which, without being indispensable as a basis of probability theory, can acquire a certain value in the investigation of the algorithmic side of mathematics as a whole"*. These words beautifully summarize his scientific path on the subject. He started using information theory in mathematical sciences to rebuild several concepts on its light and ended in rebuilding information theory through the theory of algorithms, thus obtaining a logico-algorithmic foundation of probability theory. The concept of algorithmic complexity indeed is deeply interwoven with that of entropy and, in the deterministic context, with chaos. This twist is illustrated by examples in the entry Complexity measures, dedicated to complexity and randomness.

We like to end this brief and far-from-exhaustive account on Kolmogorov contributions with a personal note of the authors of this book. In our scientific career, all of us read and used several important results of Kolmogorov. As it usually happens to the greatest scientists, most of the times, we used his results just quoting his name with no references, because everybody knows them. In particular, one of the author is so (scientifically) beholden to ANK to be brave enough to bring in the wallet ANK's portrait on the back of his wife portrait.

Chapter 18
Laplace

The French mathematician, astronomer, physicist and philosopher Pierre-Simon marquis de Laplace (1749–1827) has been among the most influential scientists in the history. He was born in Beaumount-en-Auge, Normandy on March 23 and died on March 5 in Paris, ten years after being named marquis under the Bourbon restoration.

He gave important contributions in many fields of physics and mathematics both pure and applied, also impacting philosophy. Differently from other great scientists, he is not to be remembered for some revolutionary idea but surely for the perfection he reached in all aspects of science he touched. This view was clearly expressed in his funeral eulogy by the colleague Joseph Fourier *"We cannot affirm that it was given to him to create an entirely new science as did Archimedes and Galileo; to give to mathematical theories original and immensely extended principles, as did Descartes, Newton and Leibnitz; but Laplace was born to perfect all, to penetrate everything, to extend all limits, to solve what was believed to be insoluble. He would have completed the science of the skies, if this science could be completed"*. In more recent times, Jacques Merleau-Ponty, in commenting on Laplace's scientific activity, declared *"all astronomical, cosmological, and physical work falls within a paradigm that he has above all contributed to define, extend and consolidate. He is not the author of any scientific revolution. This makes him slide toward the anonymity of the acquired truths, of the immovable certainties in their field and at their level"*. These words are actually very true considering his outstanding contributions to physics and mathematics and the fact that he is practically unknown outside the limited circle of scientists and specialists.

Laplace came from a modest family. His father, Pierre, was a cider merchant and his mother, Marie-Anne Sochon, came from a small land-owning family. The early education of Laplace was at the Benedictine college in Beaumount-en-Auge from the age of 7 to 16. Laplace's father had thought of an ecclesiastical career for him and at the age of 17, as a cleric, the young Pierre-Simon enrolled at the College of Art at Caen, a Jesuit school, with the intention of studying theology. During the first years of studies he discovered his mathematical gifts. Legend tells that one day he

© Springer Nature Switzerland AG 2021
M. Cencini et al., *A Random Walk in Physics*,
https://doi.org/10.1007/978-3-030-72531-0_18

found a book on advanced mathematics that he read with avidity, and from that day his vocation was fixed. In 1768, he abandoned his theology studies and left Caen for Paris, with a recommendation letter of his mathematical teacher at Caen directed to Jean le Rond d'Alembert, member of the *Académie des Sciences* (the French Academy of Sciences) and, at the time, one of the most important mathematicians on the international scene. d'Alembert was not impressed by the recommendations but was convinced by the singular profundity of some writing of Laplace on general principles of mechanics. In d'Alembert words reported in Fourier's eulogy for Laplace: *"Sir, you see that I attach little importance to recommendations; you have no need for them; you have made yourself better known; this satisfies me; my support is due you"*. As a result, d'Alembert pledged to help Laplace and found him a position as a teacher of mathematics at the military academy. He was only 19 years old! From that moment, despite a not really exciting job (teaching mathematical analysis, geometry, and statistics to the young cadets of the *École Militaire*) he began to write an impressive number of scientific essays, and, in 1773 at only 24 years of age, he entered the prestigious *Académie des Sciences*.

Laplace's work, although best known for celestial mechanics (for his activity in this field he took the title of French Newton) and probability theory (he can surely be considered one of the fathers of modern probability theory), impacted many fields of the science of his time including, to name but a few, matrix calculation methods (in 1772 Laplace gave the first general method of computation of determinant of a matrix), partial differential equations (with the solution of linear partial differential equations of the second order with applications on sound and heat propagation), and the study on specific heat of bodies and combustion with his colleague Lavoisier. About this last field, it is worth citing the important work of 1783 with Lavoisier in which they synthesized water from hydrogen and oxygen, showing that water is not a pure element and definitively undermining the theory of the four elements (earth, water, fire, and air, common to all the ancient culture, sometimes with small modifications, e.g., wood instead of earth in Chinese culture). Moreover, other fundamental contributions due to Laplace concerned the development of powerful scientific investigation methodologies as potential theory, generating functions, Laplace transform, that today constitute the basic mathematical tools employed in many scientific contexts.

The term "celestial mechanics" was actually introduced by Laplace, and this field surely comprises among the most important contributions of Laplace to science. His remarkable dedication to this discipline was very well expressed by Eric Temple Bell in his book *Men of Mathematics*: *"Laplace is the great example of the wisdom of directing all of one's efforts to a single central objective worthy of the best that a man has in him"*. While Newton's theory of gravitation was more than 60 years old at the birth of Laplace, much remained to be clarified. For instance, the problem of the stability of the Solar System was still an open problem: while a gravitational system consisting of two bodies can be easily solved, leading to exact previsions about the movement of the two bodies and certainty on their mutual stability, adding only one other body (even small) leads to equations for which an exact solution cannot be found. This is the notorious three-body problem (see entries Poincaré and

Chaos), e.g., Sun, Earth, and the Moon. Even Newton, reasoning about the difficulty to establish the stability of the Solar System, came to say *"the wondrous disposition of the Sun, the planets and the comets, can only be the work of an all-powerful and intelligent Being"*.

The first scientist who began to study how small bodies (for example, comets or satellites) can influence the motion of two planets was Clairaut, who introduced the perturbation theory. Although owing to the work of Euler, Poisson, and overall Lagrange, Laplace was the first to succeed in proving that the Solar System is stable. More precisely, Laplace was able to demonstrate that the semi-major axes of the orbits of the planets had no long-term variations, and their eccentricities and inclinations showed only small variations which did not allow the orbits to intersect and planets to collide. However, it is necessary to specify that the work of Laplace (and Lagrange) concerns only linear approximations of the average motion of planets, and, in general, aside from this limit, the stability of Solar System still represents a field of research.[1] Anyway, the mathematical methods (first of all, potential theory and perturbation theory) developed by Laplace to obtain the approximated stability of the Solar System must be counted among the pillars of nineteenth-century physics. All the results on the stability of the Solar System as well as many others were summarized by Laplace in his monumental treatise *Traité de mécanique céleste* (Treatise of celestial mechanics) appeared in five volumes between 1798 and 1827. In his treatise, he presented a complete and systematic compendium of the works on Newton's equations of many great mathematicians and presented his solution to numerous open problems, including the solution of the tidal problem. He offered a complete mechanical description of the Solar System by devising methods for calculating the motions of the planets and their satellites, but, unfortunately, the matter were presented as if it had been obtained entirely by him, and it is basically impossible to discern the important contributions of other scientists, known to have previously solved some of the presented problems.

We cannot resist to the temptation to quote a famous anecdote about the *Traité de mécanique céleste*. After the publication of the monumental treatise, Napoléon Bonaparte wanted to make fun of Laplace saying that he had written a huge book on the system of the universe without mentioning God. Laplace replied with these words *"Sire, I had no need of that hypothesis"*.

One of the most impressive examples of the power of the methods introduced by Laplace in his work on celestial mechanics is surely the development of the tools (together with Lagrange) that lead to the discovery of Neptune that was observed with a telescope only in 1846 by Johann Galle. The discovery of Neptune had been possible thanks to the work of John Couch Adams and Urban Le Verrier who analyzed the irregularities in the Uranus orbit and used mathematical computations based on Laplace work. The prediction in the Neptune position was accurate within a degree!

---

[1]In particular, recent numerical simulations established that the motion of planets in the Solar System is chaotic (see entry Chaos), and therefore any prediction of long-term stability (for times larger than a few million years) is meaningless.

As said earlier, Laplace is also very well known for his works on probability theory (see also the entry Probability). It should not surprise thus that he also made use of probability in celestial mechanics. In particular, the analysis of comets' trajectories (aimed to discriminate whether the variety of the trajectories of comets are the result of chance or not) was one of the very first examples of applications of probability theory to astronomy. Laplace was the first scientist to firmly believe in the applicability of probabilistic methods in contexts different from games of chance and hypothetical urns, which were the restricted areas where the theory of probability had been confined till that time. It was thanks to Laplace that the term "probability" itself acquires the broad meaning it has today, having extended it to more general fields as applied sciences, statistical inference, estimation of scientific error, and philosophic causality. An illuminating paragraph on Laplace's ideas about determinism, chance, and on the use of probability in a scientific context is the following: *"All events, even those which by their smallness and their irregularity do not seem to belong to the general system of nature, are a consequence as necessary as the revolutions of the Sun. We attribute them to chance, because we ignore the causes which produce them and the laws which link them to the big phenomena of the universe; thus the appearance and movement of comets, which we know today to depend on the same law that brings the seasons, were previously thought of as the effect of chance by those who arranged these stars amongst the meteors. Thus, the word chance expresses only our ignorance on the causes of phenomena which are seen by us to happen and repeat without any apparent order and regularity. Probability is relative in part to this ignorance, in part to our knowledge"* (1786). In later works, Laplace will return to the problem of determinism and chance but his ideas will not differ much from those already specified in the above sentence: for Laplace all phenomena are intrinsically deterministic, probability is only a mean imposed by our inability in controlling all the conditions determining a given phenomenon.

Laplace wrote a fundamental essay in theory of probability, *Mémoire sur la probabilité des causes par les événements* (1774), in which he determined the probability of the causes, given the knowledge of the events. Uncertainty concerns both events and their causes and, in the mentioned work, Laplace used the mathematics of probability as a tool to understand the causes that determine some particular effects. This is nothing but the Bayesian approach to probability, 11 years after Bayes' famous theorem was published. It seems reasonable that, at that time, Laplace did not know Bayes' work (appeared in *Philosophical Transactions of the Royal Society of London* in 1763, 2 years after the death of Thomas Bayes), which he quoted only in later works. Anyway, we can surely state that Laplace refined and deepened Bayes' theorem over the years.[2] Moreover, Laplace emerged for the use he made of Bayes' theory: already in the first work of 1774 he gave an example of a practical application for the minimization of observational and instrumental error in astronomical

---

[2]It is important to mention his *Mémoire sur les probabilités* (1781) in which, after giving the definition of conditional probability, he gave the proof of Bayes' theorem. And *Mémoire sur les approximations des formules qui sont fonctions de trés grands nombres* (1786) in which he presented a refinement of Bayes' theorem for the case of discrete events, assuming a uniform prior.

observation. This early use of probability in an astronomical context was not an accident, as Laplace began to trace a scientific path that he would never abandon during his entire career, i.e., the use of probabilistic methods in astronomy and science, in general, including also social sciences.

Laplace activities in probability theory culminated in the celebrated treatise *Théorie analytique des probabilités* (Analytic Theory of Probabilities), first published in 1812, in which he presented in a more consistent and systematic way all the important results already discussed in many previous works and some new applications of the theory. This was an ambitious and influential book which remained a reference for the following decades as *"The mont Blanc of mathematical analysis"*, in the words of De Morgan. The book also contains two derivations (previously obtained by Laplace in 1810) of the Central Limit Theorem,[3] one of which is a very general proof that considered the sum of independent random variables. The theory was then applied not only to the ordinary problems of games of chance but also to the inquiry into the causes of phenomena, the problem of measurement error, the actuarial problem of duration of life, and many other problems till getting to the last paragraph entitled "On moral expectation". This book is undoubtedly the fundamental legacy of Laplace to the theory of probability.

It is worth remarking here that the first edition of *Théorie analytique des probabilités* contains a paragraph which can be considered the birth of Laplace's demon: *"All events, even those that on account of their smallness seem not to obey the great laws of nature, are as necessary a consequence of these laws as the revolutions of the sun. An intelligence which at a given instant would know all the forces that move matter, as well as the position and speed of each of its molecules; if on the other hand it was so vast as to analyze these data, it would contain in the same formula, the motion of the largest celestial bodies and that of the lightest atom. For such an intelligence, nothing would be irregular, and the curve described by a simple air or vapor molecule, would seem regulated as certainly as the orbit of the sun is for us"*. Again these words demonstrate Laplace's view of the intrinsic determinism of the natural phenomena, and their appearance in a book on probability is because our ignorance about some of the conditions determining a given phenomenon requires the use of the theory of probabilities.[4]

Laplace was also a central figure in the spreading of scientific culture. Certainly influenced by the spirit of the time and by the ideals of the Enlightenment, with many other French intellectuals, he was engaged in the writing of the "Encyclopédie", co-edited by his mentor Jean le Rond d'Alembert, with the stated purpose of "changing the way people think". Laplace was deeply interested in presenting his scientific

---

[3]One of the fundamental results of the probability theory asserting that the distribution of a sum of independent random variables can be approximated by a Gaussian distribution. The first introduction of the Central Limit Theorem together with the introduction of the Gaussian distribution was due to de Moivre in 1730, but Laplace extended and generalized the theorem to the sum of independent random variables with discrete distribution.

[4]For the discussion about causal determinism and the most famous statement on Laplace's demon presented in *Essai Philosophique sur les Probabilités* (A Philosophical Essay on Probabilities, 1814), we refer the reader to the entry on Determinism.

findings using simple, clear but never simplistic writings that, without the heavy mathematical apparatus, was perfect for the general audience. An impressive example of this activity was the *Exposition du système du monde* (1796), a magnificent text in five books in which Laplace aimed to make everyone understand the dynamics of the celestial bodies and the interactions with terrestrial phenomena, showing that every single effect can be deduced from Newton's laws. Somehow this treatise can be considered as the general audience counterpart of the Treatise on celestial mechanics. The final pages of the fifth volume present a striking intuition about the origin of the Solar System from a primitive nebular—the nebular hypothesis.[5] Remarkably, the concluding 19 pages of this Book V, containing the nebular hypotheses, have generated more comments than all Laplace's other works on celestial mechanics put together!

Another important treatise addressed to the general audience was the *Essai Philosophique sur les Probabilités* (A Philosophical Essay on Probabilities) initially inserted as an introduction to the second edition of *Théorie analytique des probabilités* in 1814, and then printed as an independent book with seven editions. This treatise is an accessible exposition to the concepts and philosophy of probability and it is based on a lecture given by Laplace at the École Normale Supérieure in 1795. In this book, Laplace first presented the principles of probability theory, then discussed a whole series of applications, ranging from games of chance to the composition of the tables of mortality up to application to moral science and the probability of testimonies, fulfilling the program announced at the outset of his lesson *"In this lesson, I will talk about probability theory, interesting theory by itself and by its many relationships with the most useful objects of society"*. Remarkably, a farseeing chapter is dedicated to the illusions in the estimation of probability, an issue that has been seriously addressed only in recent times. Finally, the book ended with a short history of probability theory, starting from Pascal and Fermat.

The *Essai philosophique sur les probabilités* has certainly had a longer life and almost certainly a larger number of readers than any other of Laplace's writings, including its counterpart in celestial mechanics "Exposition du systéme du monde". The reason of this growing success has clearly been the importance that probability, statistics, and stochastic analysis have increasingly assumed in all sciences and in the philosophy of science. The last words of the book *"It is seen in this essay that the theory of probabilities is at bottom only common sense reduced to calculus [...] if we consider again that, even in the things which cannot be submitted to calculus, it gives the surest hints which can guide us in our judgments, and that it teaches us to avoid the illusions which ofttimes confuse us, then we shall see that there is no science more worthy of our meditations, and that no more useful one could be*

---

[5]We recall that Immanuel Kant was the first who introduced the idea that the Solar System originated from nebulous material in his *Allgemeine Naturgeschichte und Theorie des Himmels* (Universal Natural History and Theory of the Heavens, 1755) but, as was Laplace custom, he did not mention Kant's previous work, even if, in this case, it is actually unlikely that Laplace knew the work of Kant. In later times, in 1821, given the general interest that this part had in a large audience, the Book V of the *Exposition du système du monde* was printed, with some modifications, as an independent volume entitled *Précis de l'historie de l'astronomie* (Summary of the history of astronomy).

incorporated in the system of public instruction" would surely not sound outdated in a modern essay on probabilities and demonstrate how modern was Laplace's vision, already more than 200 years ago.

It is important to remark that Laplace's outstanding scientific production took place in a very difficult and complicated historical period, encompassing the years of the French Revolution, the Civil War, the Napoleonic France, and the Bourbon Restoration. Years in which the course of history could easily overwhelm men in sight (as it happened to his friends and colleagues Marquis de Condorcet and Antoine Lavoiser who died in 1794, one in prison and the other guillotined). Even in this context, Laplace demonstrated uncommon skills, by adapting to the changing times and being reactive in radically changing political ideas, he always managed to preserve and increase prestigious positions. For instance, for a short period of 6 weeks starting from November 1799, Laplace was Interior Minister when Napoléon came to power. Even if Napoléon was not enthusiastic[6] and quickly removed him from this important position that did not prevent Laplace from obtaining other very prestigious offices under the reign of Napoléon: Senator and Vice President of the Senate in 1803, Count of the Empire in 1806, Order of the Reunion in 1813. Nevertheless, when in 1814 it was evident that the empire was falling, Laplace sided with the Bourbon and was elected to the French Academy in 1816 under Louis XVIII, who then named him Marquis in 1817. Although Laplace's behavior might seem political opportunism it was actually the loss of confidence in Napoléon after the debacle of Napoléon's Russian campaign to make him change political sides. Laplace itself expressed very clearly the principles of good policy: *"the practice of the eternal principles of reason, justice and humanity that produce and preserve societies, there is a great advantage to adhere to these principles, and a great inadvisability to deviate from them"* and a direct criticism to the Napoléon's ambitions: *"Every time a great power intoxicated by the love of conquest aspires to universal domination, the sense of liberty among the unjustly threatened nations breeds a coalition to which it always succumbs"*.

We conclude with his last words that further demonstrate his scientific stature *"what we know is very little, what we ignore is immense... man chases only ghosts"*.

---

[6]From the Napoléon's correspondence *"In the Interior, minister Quinette was replaced by Laplace, a first-rate mathematician, but who quickly proved to be a more than mediocre administrator. From his very first work, the officials realized they had made a mistake; Laplace never took a question from its real point of view; he looked for subtleties everywhere, only had problematic ideas, and finally brought the spirit of the infinitesimal into administration"*.

Chapter 19
Laws, Levels of Descriptions, and Models

According to Einstein *"the grand aim of all science is to cover the greatest number of empirical facts by logical deduction from the smallest number of hypotheses or axioms"*.[1] A (perhaps crude but effective) way to rephrase Einstein's sentence is that the final goal of physics is to write down a few equations describing the unified laws of our universe. The pop version of this point of view is the hope (dream?) to have, in the next decades, such fundamental laws or equations printed on T-shirts, as, for instance, the well-known $E = mc^2$ or $F = ma$.

A natural question arise: what is a physical law or (as it is often referred to) a law of nature? This is actually a very old and much debated question, which dates back to the origin of the philosophy of science. It concerns the "true character" of the laws of nature, whether they are real facts reflecting the deep structure of our world or just useful tools (models) that allow us to describe and (sometimes) to predict or control certain phenomena. As one can imagine, this is a quite slippery and subtle issue that could lead us too far. In the following, we shall thus limit our considerations to the working activity of scientists and to the relationships between different laws and different levels of description of the physical reality.

The followers of the reductionist approach believe that, at least in principle, one can explain all the phenomena in terms of fundamental laws involving elementary constituents, as the processes at atomic (or quarks, or strings) level.[2] According to this point of view, this is the way to reach a deep insight into the understanding of nature. Such an opinion has been clearly expressed by Steven Weinberg[3]: *"we*

---

[1] As for many quotations of Einstein, it is not always easy to be sure of their authenticity; however, in such a case we believe that this sentence surely reflects the opinion of the great scientist.

[2] In the extreme version of the reductionism, the psychology can be reduced to biology, which is nothing but a part of chemistry, which can be reduced to quantum mechanics, and so on until quarks and strings, see the entry Determinism.

[3] Nobel Prize in physics (1979) for his contributions to the standard model and champion of the reductionism.

© Springer Nature Switzerland AG 2021
M. Cencini et al., *A Random Walk in Physics*,
https://doi.org/10.1007/978-3-030-72531-0_19

understand perfectly well that hydrodynamics and thermodynamics are what they are because of the principles of microscopic physics".

A natural criticism to this point of view consists in noticing that claiming "in principle it is possible" without showing an explicit procedure sounds as a dogmatic statement, and thus it cannot be considered satisfactory. On a more technical ground, physics offers a plenty of examples, some of which are briefly discussed in the entry Universality, showing that it is not always necessary to have a perfect control of the microscopic level to master and understand the macroscopic behavior of a system.

With no intention to lessen the role of reductionism in physics and more generally in science (subject which often has given rise to hot debates), we cannot avoid mentioning Phil Anderson[4] who defends positions very different from Weinberg's ones. For instance, he argues that: *"The ability to reduce everything to simple fundamental laws does not imply the ability to start from those laws and reconstruct the universe"*. Such a point of view is quite interesting here as it is expedient to consider concepts as physical theories and laws from a perspective closer to the practical scientific activity. Indeed, looking both at the past and present researches in physics, in our opinion, there is no clear evidence that the research of ultimate levels along the path "macroscopic objects $\rightarrow$ atoms $\rightarrow$ elementary particles $\rightarrow$ quarks $\rightarrow$ string" represents the only intriguing and worth pursuing scientific activity. Even if a final global theory of the fundamental laws (the grand unification) will be discovered, undoubtedly it remains a lot of work to rationalize mesoscale and macroscale natural phenomena. Such an opinion was clearly expressed by Richard Feynman: *"The age in which we live[5] is the age in which we are discovering the fundamental laws of nature, and that day will never come again. It is very exciting, it is marvelous, but this excitement will have to go. Of course in the future there will be other interests. There will be the interest of the connection of one level of phenomena to another phenomena in biology and so on, or, if you are talking about exploration, exploring other planets, but there will not still be the same things that we are doing now"*.

To illustrate the difficulty to "start from [the fundamental] laws and reconstruct the universe", it is instructive to consider two specific cases.

As a simple and well-known example of the different levels of descriptions and effective theories which can be used while considering a physical phenomena, we briefly consider the case of optics. Since the beginning of the twentieth century, with the work of Planck and Einstein, we know that light can be described in terms of photons and quantum mechanics. This approach is the only one able to account for the way light and matter interact with each other. On the other hand, for accounting light interference, diffraction, polarization, and several other phenomena happening at length scale much larger than the atomic scale, there is no need of using quantum optics and we can use physical or wave optics, that is, Maxwell's equations of elec-

---

[4]Nobel Prize in physics (1977) for his contribution to the theory of condensed matter, after his celebrated paper *More Is Different*, Science **177**, 393 (1972). He is considered the great patron of the anti-reductionist point of view.

[5]Here Feynman, Nobel Prize in physics (1965) for his work on quantum electrodynamics, is referring to the golden age of particle physics.

tromagnetism. Even this last level can be too accurate in our daily life, for instance, we use the term shadow, but strictly speaking shadows do not exist in wave optics. The effective theory used in our common activity outside scientific laboratories, e.g., by the opticians, is the so-called geometrical (or ray) optics which is valid if the typical linear size of the body is much larger than the wavelength of the light. Only in such limit it is possible to speak of shadows, i.e., a sharp distinction darkness/light. Can you imagine if to manufacture glass lenses one would need to know quantum optics?

We now consider and elaborate on the case of a complex geophysical phenomenon as El Niño, namely the nearly (but not exactly) periodic variation of the sea temperature and winds in the tropical eastern Pacific Ocean, which influences most of the climate around tropics and sub-tropics, with important economical consequences. Of course, nobody in his right mind would really follow a rigid reductionist approach for such a difficult problem, i.e., to derive the equations used by geophysicists with a procedure which starts from the microscopic level. Because of the gigantic difference between the characteristic times of the phenomena at atomic level (say 10^{-12} s) and those at geophysical level (several months), we cannot do the direct jump

$$\text{microscopic level (molecules)} \rightarrow \text{effective equations for El Niño}$$

but a more articulated route, which can be outlined as follows:

$$\text{molecular level} \rightarrow \text{fluid dynamics} \rightarrow$$
$$\text{quasi-geostrofic equations} \rightarrow \text{effective equations for El Niño,}$$

where what is on the left of "$\rightarrow$" corresponds to "is more refined, or requires more resolution than" and is not intended as "can be easily used to deduce the laws of" what is on the right.

In the entry Richardson, we briefly discussed the path from the fluid dynamics equations to effective equations relevant to weather phenomena, which roughly corresponds to the step "fluid dynamics" $\rightarrow$ "quasi-geostrofic equations". Here below, we sketch the passage from the microscopic level (avoiding considering the more detailed level of quantum mechanics) to the macroscopic level, i.e., the step "molecular level" $\rightarrow$ "fluid dynamics". In its simplest formulation, the scheme is roughly the following:

I microscopic (discrete) level: Newton's equations for the atoms (molecules) composing the fluid;

II kinetic theory (Boltzmann's equation): from a mechanical to a probabilistic description which involves two key steps, namely, coarse-graining[6] and the approximation known as *molecular chaos* to handle the collisions among molecules;

III macroscopic-level description (the Navier-Stokes equations): assuming that the length traveled by each molecule before colliding with another is much smaller than other length scales in the system, with a suitable non-trivial limit one finally obtains the fluid dynamics continuum description.

---

[6]One has to build a probabilistic description by considering all the molecules within small volumes.

Remarkably, as briefly discussed in the entry Turbulence, the above passages are not really necessary to derive the Navier-Stokes equations, as they can be directly obtained applying Newton's laws to a continuum.[7] However, one may still object that the continuum path is just an heuristic (phenomenological) approach which luckily leads to the correct result, and that the above approach is more fundamental and based on the "correct" microscopic (molecular) description of the phenomenon. Actually, it turns out that the macroscopic level (III) is not so sensitive to the fine details of the microscopic level (I). For instance, in the 1980s, it was developed an interesting approach based on probabilistic cellular automata, in which the deterministic microscopic dynamics of the fluid molecules is replaced by a discrete model defined on a lattice (the so-called lattice-gas automata). In such an artificial world, the particles have only a discrete set of velocities and hop from one site of the lattice to a nearest-neighbor one. Whenever two or more particles meet in the same site, a "collision" changes their velocities according to rules which satisfy some minimal physical requirements, namely, the conservation laws and some symmetries. Remarkably, performing an average over a spatial region much larger than the lattice spacing (i.e., a coarse-graining), one obtains a macro-dynamics which is a very good approximation of the Navier-Stokes equations. This shows that the macro-level description can be obtained by different microscopic dynamics (something similar happens in the context of critical phenomena as discussed in the entry Universality), this is due to the fact that the projection due to the coarse-graining combined with the limiting procedure wash out a lot of inessential (for the macroscopic level) microscopic details.

Figure 19.1 (next page) expands a bit the above sketch and displays the different theories required to describe the properties of fluids at different scales. The crossing from one theory (level of description) to another (from left to right) is determined by a coarse-graining and/or a projection procedure with a "loss of information". The key difference between this approach and strict reductionism is that each level of description is somehow autonomous and self-contained. In the attempt to deduce laws from one level to another, several details are lost and sometimes new information is gained: it "emerges", as many say (see the entry Irreversibility).

It is important to emphasize that, unlike the widespread belief, the limiting procedure permitting the passage from a theory (level of description) to another is very far to be a mere approximation as it typically involves mathematically delicate (often) singular limits. In this respect, we mention that, even in good textbooks, one can find the wrong claim that classical mechanics is nothing but a limit of the quantum mechanics. This would mean that quantum mechanics is more fundamental than classical mechanics, in the sense that the latter theory is just an approximation of the quantum mechanical description. One of the father of quantum mechanics (Heisenberg) expressed, in a rather clear way, a very different opinion: *"Newtonian mechanics is a kind of a priori for quantum theory. It is a priori in the sense that it is the language which enables us to say what we observe"*. A rather similar point of view can be found in the influential textbook of Landau and Lifshitz: *"It is in prin-*

---

[7]This is actually the path which was followed by Euler and then refined by Navier and Stokes.

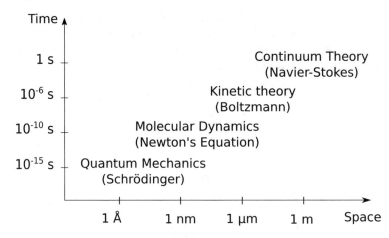

Fig. 19.1 The different theories used at different scales (description levels) in the study of a fluid

ciple impossible [...] to formulate the basic concepts of quantum mechanics without using classical mechanics".

In Heisenberg's philosophy, the basic concept is that of "closed theory", i.e., a system of mathematical objects, axioms, laws, and so on describing a certain set of phenomena in a correct self-consistent fashion. Heisenberg identifies four closed theories in physics: classical mechanics, electromagnetism (including optics and special relativity), statistical mechanics (including thermodynamics), and quantum mechanics. For each of these four theories, there is a precisely formulated system of concepts and axioms, whose propositions are strictly valid within the particular realm of experience they describe. Roughly speaking, Heisenberg had a picture of theoretical physics which is not unified, but consists of non-overlapping theories, each with its own methods and range of applicability.

It is not easy to explain in a simple way the different procedures which had been followed to build the many models used in science, also because it can be safely claimed that there are not systematic protocols for model building. The necessity to use appropriate models for the different levels of description is particularly evident in all the problems characterized by the presence of very different time scales. Among the most important examples, we mention protein folding and climate. For proteins, the time scale of the vibration of covalent bonds is in the order of 10^{-12} s, while the folding time can be of the order of seconds. In the case of climate, the characteristic times of the involved processes vary from hours/days (for the atmospheric phenomena) to thousands years for the deep ocean currents and ice shields. Since the obvious impossibility to handle (even with computer simulations) all the many different variables with so different characteristic times involved in this kind of systems, we are forced to reduce our ambitions by focusing and modeling only the most relevant aspects and variables of the phenomenon we are interested in, which often refer to the slowly varying variables.

The necessity of treating the "slow dynamics" in terms of effective equations is both practical (avoid impossible numerical computations) and conceptual: effective equations are able to catch some general features and to reveal dominant ingredients which can remain hidden in the detailed description. The study of the problems involving multiple scales has a long history in science dating back to Newton, and some clever mathematical methods have been developed, whose usage, however, is often not easy in non-idealized situations.

For instance, in the entry Brownian Motion, we discussed the historical and conceptual relevance of this random behavior that represents one of the first examples of multiscale physical modeling. In their studies, Einstein, Smoluchowski, and Langevin recognized the multiscale structure intrinsic in the movement of a particle of $\sim 10^{-6}$ m size suspended in a fluid made of molecules of $\sim 10^{-10}$ m size, and exploited such a structure to achieve an effective theoretical model, with a modern terminology a Gaussian diffusive process. A similar situation is encountered when considering the problem of particles transported by a fluid flow, one can show that at very large times and scales all the details of the velocity field become inessential and one can use an effective equation which is basically diffusion (see the entry Turbulence) but with a diffusion coefficient much larger than that of Brownian Motion (which account for molecular effects) as it includes in an "effective" way the contribution of the velocity field (in technical jargon this is called the renormalized or turbulent diffusion coefficient). Hence, the finer details of the velocity field of the fluid are washed out and "summarized" in a constant[8]—the diffusion coefficient. Whenever one is interested in describing the transport of substances on very long time and large scales, the use of such an effective description is much more useful and handy than considering the complete problem of fluid transport and molecular diffusion, which are anyway high-level descriptions with respect to the microscopic ones.

In conclusions, we can say that models are representations and/or abstractions of real phenomena that allow for predictions of some aspects of the phenomena and their understanding at a certain level of resolution. In some sense, even those very complete and elegant descriptions that we call theories, such as classical and quantum mechanics or electrodynamics, are nothing but very sophisticated models.

---

[8]Or a few constants if the flow is not isotropic.

Chapter 20
Maxwell

James Clerk Maxwell (1831–1879) had been one of the greatest physicists of history, his contributions can be considered to be of the same levels of those of Newton and Einstein. About him Einstein wrote *"One scientific epoch ended and another began with James Clerk Maxwell"*, and Feymann *"From a long view of the history of mankind—seen from, say, ten thousand years from now—there can be little doubt that the most significant event of the nineteenth century will be judged as Maxwell's discovery of the laws of electrodynamics"*.

His seminal works on the kinetic theory of gases and the electromagnetism are well known, but he also gave important contributions to other fields, e.g., color vision, celestial mechanics, and technology, in particular, on instruments based on light and also steam engines.

Although he is known as Maxwell, the original family name was Clerk: his father John, for some reason related to an estate, added Maxwell to the family name.

James was born in Edinburgh, his family was financially well off and well established, he grew up at Glenlair, in the family estate.

In 1847, James enrolled at Edinburgh University, where he studied mathematics and logic; in 1850 he went up to Cambridge. In 1854, he graduated as Second Wrangler in the Tripos exam (i.e., second in the honors degree course) and then he was joint winner of prestigious Smith's Prize.

At the beginning of his post-graduate activity, his research was mainly on experiments about the perception of colors. He quantified the effect of combining the three primary colors (red, green, and blue) distinguishing between the results of mixing lights of different colors and mixing pigments. In particular, he confirmed Helmholtz's discovery that, at variance with a common idea, mixing blue and yellow light does not produce green, but rather "a pinkish tint". It is not well known that Maxwell, in order to show his results on the perception of colors, obtained the first color photograph.

In 1856, James Clerk Maxwell (JCM) was appointed Professor of Natural Philosophy in Aberdeen, at that time a center for academic life and rather active commercial city. The period in Aberdeen was very happy and productive for Maxwell. In 1858, he received the Adams Prize for his important work on the stability of the motion of Saturn's rings. Maxwell, using rather elegant and sophisticated mathematical methods, was able to prove that solid rings could not be stable; similarly, he also shown that the hypothesis of fluid rings must be rejected. The conclusion was that the unique possibility in agreement with the stability is that each ring consists of different separate parts. Thanks to Voyager mission of the 1980s, we now know that Saturn's rings consist of ice and rocky material, ranging from micrometers up to meters. This has been the unique work of JCM on celestial mechanics, but it cannot be considered as a mere difficult mathematical exercise in problem-solving, as it had played an important role in the development of his ideas in kinetic theory and statistical mechanics.

In 1860, the two colleges of Aberdeen merged into one university, and Maxwell remained without a permanent position. After an unsuccessful attempt for the Edinburgh Chair of Natural Philosophy, Maxwell applied for, and obtained, a vacant chair at King's College, London. For the next 6 years, JCM was very productive and divided his time between London and Glenlair.

In 1865, Maxwell resigned, apparently to return to Glenlair where he occupied his time with experiments and speculations. The time he spent at Glenlair was not a period of retirement: he wrote the textbook *Theory of Heat* (1871), the *Treatise on Electricity and Magnetism* (1873) and developed his ideas on the theory of gases introducing the so-called Maxwell's Demon, which will be discussed later.

In 1871, he was called back to Cambridge to direct the new Cavendish Laboratory of Experimental Physics. The choice of a theoretician as director of a Laboratory can appear a bit strange, actually Maxwell was not the first to be approached, before him William Thomson (also known as Lord Kelvin) had been approached first and then Helmholtz.

JCK died in Cambridge (November 1879) and was buried in Parton cemetery, near his beloved Glenlair.

Undoubtedly, Maxwell's contributions to statistical mechanics and electromagnetism are recognized as among the greatest steps of the history of science, a link between the classical physics and the revolutions of twentieth century, i.e., relativity and quantum mechanics. However, JCM had deep interests also in many other scientific topics as well as in the organization of science and teaching. We already mentioned his contribution to color vision and Saturn's rings. In the academic tradition of Cambridge, a central role for the selection of new young scientists was played by the Mathematical Trypos (honors degree course) mainly based on old style mathematics. JCL in 1865 was invited to be an examiner and he succeeded in introducing new topics, and after few years electricity, magnetism, and heat entered the program.

During the International Exhibition of 1862 in London, JCM was involved in the organization of the section "instruments connected to lights" and wrote a summary published in the volume that recorded the International Exhibition, a beautiful example of didactic writing. Maxwell gave important contributions to the Committee on the Electric Standards (1861–1863), a significant step to establish the standards for the volt, ampere, and ohm.

In addition, JCM was the first to use in an explicit way the dimensional analysis (see the entry Universality), a simple but powerful method to face scientific and technological problems.

His paper *On Governors*, about the centrifugal device used to regulate steam engines, had been quite relevant for the mechanical technology of the second half of the nineteenth century and it is now considered a seminal work in the development of feedback control theory.

The building of a theoretical understanding of thermodynamics from the microscopic dynamics started with Daniel Bernoulli who introduced (in 1738) a simple basic kinetic theory. This work was soon forgotten and we had to wait for Rudolf Clausius to find a first attempt towards a theory of gases (1857–1859). In his work, JCM established the foundation of a kinetic theory, obtaining predictions which could be experimentally tested. He started by studying collisions among elastic spheres arriving to two important results: the probability distribution of the velocity of the molecules (the celebrated Maxwell-Boltzmann distribution) and an expression for the coefficient of viscosity for gases.

With his work for the velocity distribution, JCM had been a pioneer of the use of probability theory for the investigation of natural phenomena. The great J. W. Gibbs in 1889 wrote: *"In reading Clausius, we seem to be reading mechanics; in reading Maxwell, and in most of Boltzmann's most valuable work, we seem rather to be reading in theory of probability"*.

The theoretical result of Maxwell for the viscosity, namely, its independence of the density of the gas, was at odds with intuition as well as the few rough experimental data known at that time. For this reason, JCM performed an experiment (in his London house, with the help of his wife) which confirmed his apparently "curious" prediction. Such an experimental result gave strong momentum in favor of the kinetic theory.

Let us mention another celebrated contribution of Maxwell to statistical mechanics: his famous "demon" (see also entries Entropy and Irreversibility). JCM introduced the demon with the aim to clarify the fact that the second law of thermodynamics does not have a mechanical character but it is a statistical law: *"[...] if we conceive of a being whose faculties are so sharpened that he can follow every molecule in its course, such a being, whose attributes are as essentially finite as our own, would be able to do what is impossible to us. For we have seen that molecules in a vessel full of air at uniform temperature are moving with velocities by no means uniform, though the mean velocity of any great number of them, arbitrarily selected, is almost exactly uniform. Now let us suppose that such a vessel is divided into two portions, A and B, by a division in which there is a small hole, and that a being, who can see the individual molecules, opens and closes this hole, so as to allow only the swifter molecules to pass from A to B, and only the slower molecules to pass from B to A. He will thus, without expenditure of work, raise the temperature of B and lower that of A, in contradiction to the second law of thermodynamics"*.

The term demon (its first appearance was *"intelligent demon"*) was introduced by William Thomson. Although Maxwell wrote explicitly that the important point of the demon argument was to show that the second law of thermodynamics has just a statistical nature which can be violated only in small systems, there was a rather vivid (often improper) interest for Maxwell's demon even beyond physics. Likely this was mainly due to the catchy name. Remarkably Maxwell, who was a devote christian, did not like to trim the science to oblige theology, for him the demon was just a "valve".

Assuming that the demon must obey to the physical laws, Feynman, in his celebrated book *The Feynman lectures on physics*, shows that the mechanism cannot work: it is not possible to obtain work from thermal fluctuations in equilibrium. On the other hand, he proposed that fluctuations can be exploited when the system is out of equilibrium, this mechanism was named Brownian ratchet. Remarkably, recently, this ratched mechanism has been shown to play an important role in biophysical phenomena, e.g., molecular motors.

Electromagnetic theory is one of the topics which dominated Maxwell's research: his first publication in this field was *On Faraday's Lines of Forces* (1856), the masterpiece *Treatise on Electricity and Magnetism* was published 18 years later and accomplished his gigantic efforts to give a precise description of electromagnetic phenomena. We can briefly summarize the essence of Maxwell's research in three points.

The first one was the (successful) effort to provide an unified theoretical framework for the many phenomena of electricity and magnetism accumulated in the nineteenth century. Just to cite the main actors before Maxwell, we can remind Volta, Ørsted, Ampère, and Faraday.

The second one was to unify the electromagnetic phenomena through the approach developed by Faraday. Maxwell was able to expand the notion of line of force, introduced by Faraday, into the idea of field. Such an approach has been central not only to electromagnetism but also to many other fields, e.g., Einstein pointed out that Maxwell's work on field established the necessary basis for the relativity theory.

Finally, Maxwell used his unified framework to go beyond the already well-established electromagnetic phenomena. It is retrospectively impressive that JCM introduced, in his famous equations, the displacement current just on theoretical ground, without any experimental input. The introduction of such term allows for a complete and coherent structure of the mathematical description. From his equations, Maxwell was able to find, just with paper and pencil, the electromagnetic waves which were generated and detected, only after some decades, by H. Hertz (1888).

JCM derived the velocity of the electromagnetic waves (at that time not yet observed) arriving to an astonishing remark: *"The velocity of .. undulations [waves] in our hypothetical medium... agree so exactly with the velocity of light calculated from optical experiments of M. Fizeau, that we can scarcely avoid the inference that light consists in the ... undulations of same medium which is the cause of electric and magnetic phenomena"*. A first step in the direction of the unification of electromagnetism and optics.

We conclude with a methodological remark about Maxwell's use of considerations based only on theoretical arguments (i.e., completeness and coherence of the equations). This has been an important example of innovation with zero experimental basis, and it must be considered a paradigmatic case of theoretical innovation in physics.

Chapter 21
Mesoscale Systems

Lengths and durations are among the most important concepts in physics and they are crucial in our everyday life. It is thus useful to start by briefly bringing to mind the huge range of scales that animates the universe.

With our hands we experience things with dimensions spanning from millimeters to meters, and we perform activities which may take minutes or hours. By walking for a few hours we can cover kilometers; in the same time, we may travel by car for hundreds of kilometers. Airplanes lead us everywhere on Earth, traversing thousands of kilometers in a few hours.

Some objects in the universe, however, can be much larger than the Earth or much smaller than a millimeter. Such things escape our immediate perception and can be understood only through the use of scientific instruments. Microscopes and telescopes—in the last five centuries—expanded the limits of our senses, giving us access to length scales below the millimeter (down to the micron, i.e., a thousandth of millimeter) and above the thousands of kilometers. Thanks to the telescope, Gian Domenico Cassini at the end of the seventeenth century could observe details of the rings of Saturn, more than a billion kilometers away from Earth. Roughly in the same years Robert Hooke was the first to observe cells, looking at a tiny cut of cork through a microscope. The principle that durations follow lengths is reproduced at all scales: larger is usually slower. Things at the scale of microns—the so-called "microscale"— happen faster than seconds. For instance, diffusion of proteins through the cell body of an *E. coli* bacterium takes a thousandth of second. Astronomical events occur on time scales that range from months (recurrent passages of close objects such as our moon) to years (planetary orbits), to billions of years (lifetime of stars and planets).

In the last century, the range of accessible length/time scales widened even more. Biologists discovered DNA, a fundamental molecular structure thinner as a nanometer (a millionth of a millimeter); physicists performed experiments with atoms and sub-atomic particles well below the nano-scale; astronomers extended their sight to far galaxies, up to billions of light years from us, where a light year is the distance covered by light in a whole year, roughly ten trillion kilometers. The first gravita-

© Springer Nature Switzerland AG 2021
M. Cencini et al., *A Random Walk in Physics*,
https://doi.org/10.1007/978-3-030-72531-0_21

tional wave that was detected in 2015 came from 1.4 billions of light years, while the first photographed black hole (at the center of Virgo A galaxy) is at a distance of *only* 53 millions of light years.

A natural mathematical tool able to grasp and appreciate such a vastness of (spatial and temporal) orders of magnitude is the decimal logarithm. We can measure distances in decimal logarithms of a meter. Figure 21.1 (next page) illustrates the several orders of magnitude discussed in the above introduction. From the 0 or ground level of our spatial scale, a meter in fact, we can look up or down. Downward we see an ant at the −3 level, a cell at −6, an atom at −10, and the radius of an atomic nucleus at level −15. Looking up we see the diameter of a planet at the 6th level, a planetary distance at level 12, a light year at 16. A similar building can be raised using time scales instead of length scales, using decimal logarithms of some time unit, for instance, minutes or hours.

The objects of our direct experience lie at the ground or middle floor—the 0 level—of this conceptual skyscraper of scales. From our 0 level, we can stretch our instruments above for roughly 30 floors and below, roughly for 30 floors. Such an apparent symmetry has a reasonable explanation, it is just an illusion, an artifact of being us the subjects of scientific investigation; in fact, science is made of humans trying hard to extend their senses in both directions. There are also arguments to sustain that life requires a certain degree of complexity and that such a complexity naturally appears only at scales not too far below or above. In fact, it is difficult to imagine living atoms or living planets. Further discussing this point would lead us into philosophy or science fiction, which are out of our original scope.

The question, however, remains even if we stick to science: why do (many) physicists had and still devote their efforts to top and bottom levels of the skyscrapers, somehow far away from the scales of our everyday human experience? A possible reason is as follows. It is quite natural to assume that the main features of systems at levels −15 or +15, i.e., at very small or very large scales, are explained with just few ingredients. In sub-atomic experiments or in astronomical observations, interactions involve a few objects (quarks in the former or stars in the latter) and the real big challenge is a full understanding of the exact nature of the single interaction, which looks like a fundamental quest. Conversely, close to the 0 level, the details of the interactions are usually well understood—or sometimes they are even irrelevant—and the frontier is explaining phenomena which *emerge* because *many* simple objects interact together. Many scientists in physics (while in biology the natural scales of interest are those just around the 0th level) feel that such a complexity is less fascinating, less philosophical, than understanding some fundamental question.

However, if one shares the opinion that science is not philosophy, then immediately realizes that there are very interesting and often unexplained phenomena occurring close to the 0th level, fascinating discoveries waiting us at the middle floor, where the history of human exploration started centuries ago. Systems involving scales not too far from the 0th level are usually called "mesoscale" systems and in the following we briefly describe some of their most important general properties.

A paradigm of this kind of physics is Brownian Motion (see the related entry), observed by Robert Brown in 1827 while looking—through his microscope—at a

Fig. 21.1 A metaphorical cartoon where the floors of a skyscraper (above and below ground) represent the several orders of magnitudes spanned by length-scales (in meters) involved in natural and artificial phenomena

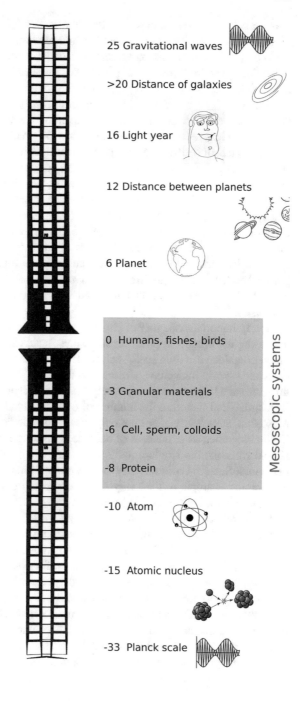

25 Gravitational waves

>20 Distance of galaxies

16 Light year

12 Distance between planets

6 Planet

0 Humans, fishes, birds

-3 Granular materials

-6 Cell, sperm, colloids

-8 Protein

Mesoscopic systems

-10 Atom

-15 Atomic nucleus

-33 Planck scale

grain of pollen suspended in water. A grain of pollen, a light particle with a diameter close to the micron, moves continuously and randomly, slowly exploring the water surface. The origin of this movement was understood by Albert Einstein (see the related entry) in his *annus mirabilis* (1905), by introducing a simple mathematical theory that predicts the basic properties of this motion, confirmed by Perrin in a precise experiment a few years later. Einstein's theory assumes that the pollen moves because it is continuously and rapidly kicked by the surrounding water molecules. Such molecules are invisible because they are too small for optical microscopes. The marvelous fact of Brownian Motion is that it bridges the very small scale of molecules (floor -10) to a much higher level, that of the pollen grain (floor $-6 \div -5$), which is accessible through a small table microscope. In fact, things are even better than that: in order to verify Einstein's theory, one needs to measure distances which are much larger, of the order of millimeters (floor -3). In conclusion, Brownian Motion is an elevator connecting floor -10 to floor -3. The elevator, however, does something quite unexpected: it transforms the laws of mechanics. These laws are perfectly followed by each single invisible molecule but are totally disobeyed by the huge grain of pollen: the strange irregular and *non-mechanic* nature of Brownian Motion is due to the complexity of the combination of billions of impacts with very simple molecules.

A very important category of physical systems at the mesoscale is living cells or, with a slightly broader application of the mesoscale concept, also the many different things contained in a living cell, such as the citoskeleton, filaments and microtubules, actin and myosin molecules, viruses, and the many different kinds of DNA (in fact, the width of a DNA strand is extremely small with respect to the cell dimensions, but its length—and therefore the number of units involved—is huge, order of meters). Mesoscale complexity arises not only inside a single cell but also when many cells interact together, as in the case of bacteria, yeasts, sperms, or even neurons. Again, simple rules for the single units lead to surprisingly rich and complex collective behaviors. Bacterial colonies may display several kinds of patterns, *Bacillus subtilis* is a shining example of this: varying the composition of its growth medium (agar and nutrients) a colony of this bacterium grows in completely different ways and under the microscope the biologist may find a disk, a sequence of concentric rings, a flower-like ("eden-like") drawing, and several types of branched tree-like images. Bacteria often have flagella that propel their movement, so that when they are in a crowded colony collective motion appears, for instance, in the form of vortices. Other typical mesoscale systems are colloids, polymers, foams, and emulsions. These last examples are not necessarily related to life, even though in many cases may be associated to organic chemistry. For instance, colloids, or, more properly speaking, colloidal suspensions, are mixtures of a dispersed phase (particles) with a continuous phase (for instance, a liquid): colloidal suspensions in everyday life are whipped cream, milk, gels, paintings, glue, and much more.

All these examples reveal the essential property of the "middle world", which is its inherent *softness*.[1] A cell, a bacterium, a vesicle, but also a long polymer, a gel or a colony of yeasts, are all prone to important deformations under external stresses. On the contrary, bodies made of ordered arrays of atoms, at the same temperature, sustain strong stresses with no deformation and, in fact, are called solids. This difference is mainly due to the fact that the building blocks of soft matter are much larger than atoms but interact with almost the same energy. Indeed units of the middle world are several floors (levels from -6 to 0) above atoms (level -10) and there is not much difference between the typical energy of interactions at the mesoscale, which is of the order of $\sim k_B T$, and that at the scale of atoms (for instance, typical energy of van der Waals forces is of the same order of magnitude) such energies are rather small.[2] Imagine to join two small objects, or two large objects, by means of the same force, what is going to happen? Intuitively one expects that the small things will stick together much more effectively than the large ones. Dimensional analysis gives an estimate for the elastic modulus (the stiffness) of a material as the ratio between energy and volume of single interacting units: this gives a ratio of 10^9 (nine levels) between the stiffness of solid with respect to that of mesoscale matter. Mesoscale matter is thus inherently soft.

Softness has many consequences: clearly it gives the advantage of a high adaptability to variations of external conditions, which is essential for the stability of life. Another consequence, which is typically a drawback, is that thermal fluctuations play a major role. Physicists call "thermal fluctuations" the fact that—at non-zero temperature $(T > 0)$—all elementary constituents of matter incessantly vibrate or fluctuate with a typical energy of the order of $k_B T$. This fluctuations affect soft matter in a dramatic way: objects inside a cell, cells themselves, colloidal units, foams, gels, etc, all these systems show several properties that fluctuate continuously. Fluctuations of their shape are perhaps the simplest to grasp. Consider a protein, which is a complex soft macromolecule composed of several solid sub-units: its shape may undergo impressive changes when temperature is varied, losing its functionalities when temperature becomes too high. Fluctuations, however, have a constructive role to increase the chances to survive. For instance, following a deterministic straight path is not an optimal strategy for a bacterium looking for nutrients: some degree of randomness is important to explore the world with effectiveness.

It is interesting to realize that, when considering the number of elementary units, mesoscale systems are *smaller* with respect to solids even if the single unit is much larger in scale (and slower in the typical time needed to perform certain processes). In fact, in a millimeter cube of mesoscale matter (e.g., a colloidal suspension or a bacterial colony), there is a number of elementary units (colloids or bacteria) which is much smaller than the number of atoms in a solid. In statistical physics, the number of

---

[1] A quite effective popularization of these concepts, including the subtle interplay between complexity and life, can be found—in fact—in the book "The middle world. The restless Heart of Matter and Life" by Mark Haw, Palgrave Macmillan, 2007.

[2] We recall that k_B is the Boltzmann constant and T is the absolute temperature in Kelvin, which means that $k_B T$ at ambient temperature is smaller than 10^{-21} calories!.

independent constituents is strictly related to the relative importance of fluctuations: when this number increases fluctuations of average properties become less and less important. In other words, we can say that many fluctuate less. To understand the origin of this phenomenon, consider the average height of children in a school: if the school is small, say 100 children, it is possible to observe an average height which is untypical, e.g., quite smaller or larger than the world average; on the contrary, if the school is large, say 1000 children, we expect to find the average height to be close to the world average. That is an alternative way of understanding why fluctuations are very important in mesoscale systems.

Artificial soft materials, i.e., mesoscale systems which are not directly related to life, pervade, with their flexibility and lightness, our world: cosmetics, food, cloths, non-stick coatings, elastics, paintings, polystyrene packaging, perspex windows, nylon pantyhose, etc. Mesoscale systems have huge applications in our everyday life, a rather important example is given by plastic. In fact, plastic is a material made of polymers. Each polymer, like proteins and DNA, is a very long chain of monomers. In the years '20s of the twentieth century the German chemist Hermann Staudinger suggested the idea that, in particular, reactions in which the formation of small circular organic molecules (such as the benzene which is a ring of six carbon atoms) could be interrupted forcing them to become longer and longer chains and that such incredibly long structures explained the properties of a lot of natural soft substances like rubber, starch, cellulose, and proteins. Industrial polymers can be extremely long, for instance, polystyrene molecules can contain millions of monomers. When a polymer is long enough, it easily takes very complex conformations because of the interplay between elastic properties of the chain and thermal fluctuations. Thanks to their flexibility, polymers can be stretched without breaking, resulting in a variety of uses. For similar reasons, polymers may tangle up and occupy a lot of space, or equivalently they do not compactify as metals and atomic solids do, resulting in very light materials.

Also in mathematics and theoretical physics, soft matter is a source of interesting problems and stimulating challenges. For instance, polymers (such as proteins or DNA) have been the ground of theoretical investigations for many years, representing a clear example of non-trivial competition between energy and entropy. Under the combined effect of internal mechanical forces and thermal fluctuations due to the environment, a long chain of monomers may assume very different shapes, more or less ordered (and thus it is characterized by a large entropy, see the related entry). A polymer also interacts with its solvent (e.g., the water where it is dispersed): the interaction between its monomers competes with the interaction with solvent molecules and thermal fluctuations, leading to cases where the polymer denatures into a long unrolled ribbon (for instance, when the solvent attracts the polymer monomers), and other cases where it becomes a very compact ball (e.g., when there is strong repulsion between its monomers and the solvent). The French physicist Pierre-Gilles de Gennes (1932–2007), who is considered one of the fathers of the subfield of soft matter, won the Nobel Prize for theoretical researches on this and other similar topics, all appertaining to the sphere of mesoscopic systems.

One of the frontiers of the investigation of soft matter and mesoscale systems is the study of self-assembly. Nature frequently demonstrates phenomena where many pieces from a random configuration *spontaneously* come together into an ordered array, displaying some particular pattern. Paradigmatic examples of self-assembly are cells that self-assemble into embryonic tissues that further develop into fully formed organs, and thus animals. Is it possible to exploit such a tendency to control an artificial pattern? Can we obtain complex materials with the desired properties which spontaneously assemble with minimal external control? This is like assembling a jigsaw puzzle without following a picture or a blueprint. The analogy with a jigsaw is fascinating: imagine a puzzle the shapes of whose pieces are designed in such a way that any piece will only fit in its proper place. Any random procedure should lead— possibly in quite a long time—to the correct final picture, without any intelligent control. The only information is in the shapes of the pieces. In biology, examples of this process are common. For example, it is common to obtain a full formed virus by a spontaneous process in a test tube that initially contains only certain proteins and RNA molecules. Assembling a puzzle in a random spontaneous way requires strategies which are totally different from supervisioned assembling. For instance, in self-assembly, thermal fluctuations play a twofold role: they favor the process by avoiding dead ends (e.g., wrong stable configurations), but they hinder the process by multiplying the false routes.

After this discussion, we may have the feeling that mesoscale systems actually exist just below the 0 level, but not above it. Actually, this is not true. Interestingly many systems of dimensions above the scale of millimeters exist and constitute stimulating challenges in the physics of mesosystems. A well-known example is that of granular matter, i.e., matter made of grains. Instances of this—which many physicists consider as a new state of matter—are sand, sugar, flour, rice, powders, large assemblies of pills, etc.[3] Understanding the behavior of granular materials is important for several industrial problems, such as the improvement of powder beds for chemical reactions or making powder mixing (or even demixing!) processes more efficient. Granular materials are also involved in many geophysical phenomena, typically related to some kind of hazard, like snow avalanches or earthquakes. Basically every kind of soil is made of (more or less compact and solid, wet, or dry) granular materials. Extraction of natural resources, such as oil or gases, requires a deep knowledge of the physics and mechanics of granular materials. Granular media constitute a perfect example of mesoscale system, as it shares several of the properties discussed above. For instance, a granular system is typically made of a small number (as compared to the atoms in a piece of solid) of constituents and randomness plays a crucial role in its behavior. Moreover, in certain earthquakes, one may observe what is called soil fluidization: an apparently solid soil takes life because of the violent shaking and becomes extremely soft and shape-changing. For some time, the vibra-

---

[3] A brief and effective popular article on the physics of sand appeared recently on the New York Times, signed by Randall Munroe also known as *xkcd*, author of thousands of comic strips, and of a few science books for a large audience, see the web page https://www.nytimes.com/2020/11/ 09/science/what-makes-sand-soft.html on the New York Times website, or the printed version on November 10, 2020, Section D, Page 4 of the New York edition.

tion acts in a similar way to thermal fluctuations and creates a sort of macroscopic soft matter. In laboratory and in theoretical studies, this phenomenon is frequently reproduced to study the differences between the soft matter made of molecules and that made of grains. Scientists find similarities and differences, and often both are surprising.

A second example of mesoscale physics at scales larger than the micron (sometimes much larger) involves the so-called "active matter". Active systems are those systems made of single units which have some kind of self-propulsion. Bacteria are a typical example. We already mentioned the case of bacterial colonies: under certain circumstance they can be considered a kind of soft matter, with small number of units, low stiffness and very important fluctuations, several kinds of ordered patterns, etc. But scientists count as active systems also the collective behavior of insects, fishes, and birds. And sometimes even pedestrians or cars! In the perspective of reducing a system to its essential features, it makes sense to put in the same category all groups of self-propelled units. It is not strange to find common phenomena belonging to very different systems made of units sharing a very important property such as self-propulsion (see the entry Universality). In fact, certain theoretical models have been used to describe both bacteria and birds. Ignoring many details, the same models can also describe granular materials, ferromagnetism, and liquid crystals.

Our opinion is that popularization of physics has been too often attracted by the two extremes of the skyscraper of scales, i.e., the top and bottom levels. In part this is a consequence of this attraction being shared by many members of the scientific community. In part this is due to the expectation of something profound and philosophical or something that answers to the ultimate question of life, the universe, and everything[4] from science. In a certain measure, science is often replaced by science fiction or misrepresented as that, even by good writers of science popularization. In the best cases, instead of science fiction, one has philosophy, see, for instance, the many books that popularize the problems of interpretation of quantum mechanics or general relativity. On the contrary, scientific investigation finds fascinating challenges, beautiful phenomena, and complicate brain teasers or puzzles, even by looking at the world close to us, in the skies where flocks fly or under our feet, while walking on a warm beach.

---

[4]We recall in this perspective that such an answer is known to be "42", as from *The Hitchhiker's Guide to the Galaxy* by Douglas Adams.

Chapter 22
Poincaré

Henri Poincaré (1854–1912) had been a scientist with a wide range of interests, mainly he was mathematician but also theoretical physicist, engineer, and philosopher of science. He can be considered one of the last universalists, since he was able to give relevant contributions in almost all fields of mathematics as well in theoretical physics and philosophy of science.

He was born in 1854 in Nancy into an influential family. His father Leon was a surgeon and professor of medicine at the University of Nancy; his uncle Antoni (Leon's elder brother), studied at Polytechnic School and became a mining engineer; his cousin, Raymond Poincaré (Antoni's son) was a lawyer, fellow member of the Académie française, and an important politician, he was several times Minister and President of France from 1913 to 1920.

In 1873, Poincaré, after a very severe selection, enrolled in the prestigious École Polytechnique in Paris; in November 1875, Henri continued his higher education at the École des Mines in Paris. Today such a path can sound rather unlikely.

From April to November 1879, Poincaré worked as inspector of mines at Vesoul, in eastern France. As a mining engineer, his job was not an untroubled activity to do in an office. For instance, a few hours after an explosion (September 1, 1879) at a depth of 650 meters with 16 killed mine workers, Poincaré descended into the mine for his investigation about the mine ventilation and gallery system.

On December 1879, Poincaré was appointed as a Lecturer in Mathematics at the University of Caen. From 1881, he taught at the University of Paris, where, in roughly 30 years, he held several chairs (Physical and Experimental Mechanics, Mathematical Physics, Probability Theory, Celestial Mechanics and Astronomy). He died in 1912 from an embolism after a surgery for a prostate problem.

It is pretty impossible to discuss in a few pages the many contributions of Poincaré as mathematician, physicist, and philosopher, we can just mention his researches on differential equations, analytic functions, number theory, topology, celestial mechanics, probability, geodesy, applied physics, and relativity. In the following, we will

© Springer Nature Switzerland AG 2021
M. Cencini et al., *A Random Walk in Physics*,
https://doi.org/10.1007/978-3-030-72531-0_22

focus only on a few of his contributions, namely, chaos, relativity, and some of his ideas which stimulated researches in apparently very far topics as psychology.

There is no doubt that the discovery of chaos and the first contribution to dynamical systems are from Poincaré's pioneering studies on the three-body problem. The discovery of chaos is associated to an old problem of celestial mechanics: the building of a table of the positions and velocities of all celestial bodies (Sun, planets, and so on) as function of time. In order to obtain such a table (the ephemeris of the Solar System), it is necessary to solve the equations of motion of celestial bodies. The case with just the Sun and one planet, the so-called two-body problem, is completely solvable: according to the celebrated Kepler's laws, there are only periodic orbits. It seems natural to wonder about a system with just three bodies (e.g., Sun, Jupiter, and one asteroid). Such a problem attracted the interest of great mathematicians as Euler, Lagrange, and Laplace. In spite of the huge efforts, even in the simplest case, called the restricted three-body problem, where one assumes that the asteroid does not induce any feedback on the Sun and Jupiter, only few special solutions had been obtained.

In 1887, the King of Sweden and Norway Oscar II announced a competition with a prize to be awarded in 1889, on the occasion of his sixtieth birthday, for a mathematical work on the three-body problem. The members of the committee were the famous mathematicians Karl Weierstrass, Gösta Mittag-Leffler (a mathematician who had an important role in the international academic life), and Charles Hermite. Poincaré applied for the prize. He did not solve the problem as originally formulated for the prize, but understood a fundamental fact, i.e., the difficulties encountered by earlier mathematicians in predicting the evolution of the three-body system are due to an intrinsic property of the dynamics, that is, tiny differences in the initial conditions were exponentially amplified during the evolution. The president of the committee Weierstrass wrote *"This work cannot indeed be considered as furnishing the complete solution of the question proposed, but that it is nevertheless of such importance that its publication will inaugurate a new era in the history of celestial mechanics"*. Weierstrass was actually right: with his work Poincaré had been at the origin of a new field which is still fertile both in mathematics and physics as well in applications. The great idea of Poincaré was to understand that sometimes, instead of calculating a particular trajectory, it is much more important and interesting to look at the qualitative behavior of all the trajectories of the system. Let us just mention that the methods introduced by Poincaré have been used in the planning of the Genesis Discovery Mission (2001–2004) allowing for a consistent reduction of the necessary fuel.

Let us open a short digression: the first version of the paper presenting the study for which he won the prize, *Sur le problème des trois corps et les équations de la Dynamique*, contained a serious error reported by the Swedish mathematician Lars Edvard Phragmén. The paper was already published on the prestigious journal *Acta Mathematica*, but with several pretexts, Mittag-Leffler (who was the editor of the journal) was able to withdraw the copies of the journal, and it was necessary to print again the issue with the corrected version. Poincaré was asked to pay the additional costs: 3585 kronor, more than the prize (2500 kronor); just to have an idea the annual salary of a Swedish professor at that time was around 7000 kronor.

Beyond this historical curiosity, the revolutionary contribution was that, in spite of the deterministic nature of the system, its evolution can be chaotic, i.e., small perturbations in the initial state, for instance, a slight change in one body's initial position might lead to dramatic differences in the later states of the system. As discussed in the entry Chaos, for a series of reasons, the deep implication of these results did not receive the due attention for a quite long time, and only after the work of Lorenz the chaotic behavior of deterministic systems attracted the interest of a wide scientific community.

For few centuries, we had a long sequence of successes of the Newtonian mechanics in the building of the myth of "astronomical precision". In the following, we provide a concise summary. First, Newton was able to derive Kepler's laws from the gravitational law. Then we had the contributions of Laplace on the stability (actually just for a long time interval) of the Solar System and, most importantly, the discovery of the planet Neptune: in the nineteenth century, a series of observations indicated a significant deviation from the motion of Uranus' orbit foreseen by Newtonian mechanics. Adams in England and Le Verrier in France suggested that this discrepancy was not due to a deficiency in Newton's theory but to the presence of an unknown planet. Based on the laws of motion and gravitation, they calculated the location of this hypothetical planet, which was shortly after observed by Galle with a telescope.

For the first time, after hundreds of years of successes and correct predictions, the result of Poincaré showed that the "astronomical precision" could be just an illusion and that there is the risk that the Solar System could be unstable. Today, after the contribution of Kolmogorov, Arnold, and Moser (1954–1962), we know that we can be optimistic: although numerical simulations suggest a very small chaotic nature of the motion of the planets, there is a good chance that the Solar System is stable (this is briefly discussed in the entry Kolmogorov).

Usually a scientific book becomes obsolete in few decades, this is not the case of the three monumental volumes of *Méthodes Nouvelles de la Mécanique Céleste* (1892–1899) which, after more than one century, are still a source of inspiration for students and researchers, and not only in the mathematical community. We just mention the fact that the English translation of this book was from NACA (now NASA) at the beginning of its activity in the space program (1952).

The scientist whose name is associated with the theory of relativity is, of course, Albert Einstein, less known is the fact that well before Einstein, Poincaré studied the theoretical problem associated to the Lorentz' approach to electrodynamics. He had wide interests in theoretical physics, in particular, in electromagnetism, in addition, since his work at the Bureau des Longitudes, he was interested in the problem of synchronization of clocks which are at rest on Earth but are moving at different speeds relative to absolute space. The first paper of Einstein on relativity was published 3 months after Poincaré's paper *Sur la dynamique de l'électron* (1905), in which he presented a dynamical derivation of the Lorentz contraction associated to moving electrons and was arguably the first scientist to express the principle of relativity in terms of the form invariance of laws of physics. Despite many similarities, Poincaré and Einstein had very different research agendas and physical interpretations of the theory. The two scientists had been giants of science, it seems to us that it is not particularly interesting to insist too much on the debate about the priority. In 1927, Lorentz wrote: *"I considered my time transformation only as a heuristic working hypothesis. So the theory of relativity is really solely Einstein's work. And there can be no doubt that he would have conceived it even if the work of all his predecessors in the theory of this field had not been done at all. His work is in this respect independent of the previous theories"*. Of course the most important among the predecessors was Poincaré, as acknowledged by Einstein himself who in 1955 said *"Lorentz had already recognized that the transformation named after him are essential for the analysis of Maxwell's equations, and Poincaré deepened this insight still further..."*.

As discussed in the entry Irreversibility, the celebrated Poincaré recurrence theorem (which appeared in the famous paper on Acta Mathematica) had an important role in the debate about the second law of thermodynamics and atoms. It is interesting to note that the point of view of Poincaré on such topics was rather far from Boltzmann's one. Poincaré was concerned with obstacles met in the explanation of the second law of thermodynamics based on mechanistic conception as well as on the idea that everything can be reduced to the motion of atoms. He was rather sharp: *"It is in any case impossible on the basis of the present theory to carry out a mechanical derivation of the second law without specializing the initial state. It is likewise impossible to prove that the well-known velocity distribution will be reached as a stationary final state, as its discoverers Maxwell and Boltzmann wished to do"*. We can say that Poincaré, in his tribute to Boltzmann, was not generous: *"Boltzmann, who died tragically, had been teaching for a long time in Vienna; he had become known especially for his researches on the kinetic theory of gases. If the world obeys the laws of mechanics that allow us to proceed both forward and backward in time, why does it constantly tend toward uniformity without any chance that one may bring it back? This was the problem to be solved which he had devoted himself to, and not without some success"*.

Poincaré interest was mainly for science and philosophy; however, he had also an important role in the academic and cultural French life, for instance, he was in charge as president for important commissions, e.g., in the French Bureau des Longitudes and the Société Astronomique de France.

At the end of the nineteenth century, a very important French event was the Dreyfus affair. In 1894, Captain Alfred Dreyfus, a Jewish officer in the French army, was accused of high treason. There was a tremendous uproar in France, with two factions: the Dreyfusards, who called for "justice for Dreyfus", and the anti-Dreyfusards, who defended the actions of the establishment, and producing frightening outbursts of anti-semitism. Even today, the discussion of the affair means trouble in certain French army circles. When a retrial was granted, a commission of three eminent scientists, Gaston Darboux, Poincaré, and Paul Appell, was asked to report on the scientific reasoning, based on some confuse probabilistic arguments that was in fact a crucial part of the accusations. The conclusion of Poincaré was rather clear: *"I do not know whether the accused will be found guilty, but if he is, it will be on other proofs. It is impossible that such an argumentation would be seriously considered by scientifically educated people without prejudice"*. In spite of this, only in 1906 with a third trial it was established that the accusation against Dreyfus was without any foundation.

Poincaré wrote several non-technical books which were real best seller, (e.g., in 1925, *La science et l'hypothése* (1902) had already been printed in 40000 copies) well known to the general public, translated in most European languages, as well as in Japanese, and circulated widely throughout the world. Even today such books are published and are still current masterpieces of high-level popular science. Beyond the media success, these non-technical books had a rather important role in the scientific and cultural life of the first decades of the twentieth century. For instance, the young Einstein was an avid reader of Poincaré's books. *La science et l'hypothése* contains a brief discussion on Brownian Motion and its possible connection with the second law of thermodynamics, and, for sure, this was one of the origins for the interest of Einstein for the Brownian Motion.

The national and international reputation of Poincaré was so high that his life had been the object of articles in daily papers in France and U.S.A. For instance, when his book *Science et Méthode* (1908) was translated in English, the newspaper The San Francisco Sunday Call published a long article.

In 1910, Édouard Toulouse, a psychologist, wrote a book entitled *Henri Poincaré* that was not just a mere tribute to a great scientist. The interest was for the organization of Poincaré's time: usually each days he worked in a systematic way only 2 hours in the morning and 2 hours in the afternoon. He believed that the subconscious would continue working on the problem, and that his best ideas would come when he stopped concentrating on a problem, when he was actually at rest. Toulouse stated that most of the mathematicians worked from principles already established, while Poincaré started from basic principles every time. His method of thinking is well summarized as: *"Accustomed to neglecting details and to looking only at mountain tops, he went from one peak to another with surprising rapidity, and the facts he discovered, clustering around their center, were instantly and automatically pigeonholed in his memory"*.

The origin of the interest of psychologists for Poincaré was mainly originated by a section (*L'invention mathématique*) in his famous best seller *Science et Méthode*, about mathematical invention (or discovery), where Poincaré describes how sudden insight came to him in solving certain mathematical problems such as the Fuchsian functions. Such a contribution had an important role in the realm of science as well in psychology, in particular, after the book of J. Hadamard, *An Essay on the Psychology of Invention in the Mathematical Field* (1945).

In his researches, Poincaré used an approach which was not particularly systematic, he was a sort of bee flying from flower to flower, without taking too long to clarify the solution of a given topic that he considered clear enough. This attitude was opposite to those of Russell and Frege, who believed that mathematics was a branch of logic. According to Poincaré, intuition and rigor must have a rather different role: the first for finding ideas, the second for establishing those ideas: *"It is by logic we prove, it is by intuition that we invent..... Logic, therefore, remains barren unless fertilized by intuition".*

Poincaré had not founded his own school since he never had any student, but, paradoxically, he inspired the development of the mathematics in the twentieth century toward a direction opposite to his ideas. The Bourbaki group, with its obsession about mathematical rigor, could be somehow considered a reaction of the French mathematicians to the non-systematic Poincaré's approach. Surely the Bourbaki group has had great merits but their point of view, which privileges the rigor and the general mathematical structures, has also had some negative aspects. For instance, important fields, such as geometry and probability, remained outside of the Bourbaki mathematical building. The famous mathematician Arnold, a great supporter of the point of view of Poincaré, in clear disagreement with the hyper rigor of the Bourbaki, wrote *"Mathematics is a part of physics. Physics is an experimental science, a part of natural science. Mathematics is the part of physics where experiments are cheap".* Another great mathematician, Thom, in explicit disagreement with the Bourbaki group, joked about the fact that the term "rigor" reminds him of the "rigor mortis".

We like to conclude this entry with a famous quote from *Science et Méthode*, which we think expresses the sentiment of most scientists, at least of the authors of this book: *"The scientist does not study nature because it is useful to do so. He studies it because he takes pleasure in it, and he takes pleasure in it because it is beautiful. If nature were not beautiful it would not be worth knowing, and life would not be worth living. I am not speaking, of course, of the beauty which strikes the senses, of the beauty of qualities and appearances. I am far from despising this, but it has nothing to do with science. What I mean is that more intimate beauty which comes from the harmonious order of its parts, and which a pure intelligence can grasp".*

Chapter 23
Prediction

According to the neuroscientific theory known as *predictive processing* or *coding*, our brains are designed to constantly generate and update mental models of the environment so as to obtain predictions. The predictions are then compared with the sensory inputs from the environment so as to evaluate the prediction error which is in turn used to improve the model. The advantage of such brain function does not need much discussion and we can test it in many daily circumstances in which we make short-term predictions, e.g., deciding if we can cross a street when car is arriving. This suggests that the need to make predictions is somehow hardwired in the way we function. Naturally, as our societies increased in complexity, the need to make also longer term predictions started to be important, e.g., when deciding if to establish a community in a place or another. It is thus unsurprising that the wish to perform a prediction of the future has been a constant aspiration in human history. In the prescientific time, we can mention the Chaldean astrologers with their horoscopes, and the Etruscan haruspices who were claiming to be able to read the future in the livers of the sacrificed animals.

Nowadays, we use scientific knowledge in attempting a prediction. Actually the ability to make predictions is a distinctive characteristic of the scientific method typical of physics and other sciences: observe a phenomenon; build a model (a theory) to explain the phenomenon; use the model to make predictions; evaluate the prediction error with experiments; and thus validate, refuse, or refine the model. Somehow we can think of science as a culturally codified evolution of brain predictive processing. However, in spite of a few centuries of expansion of the scientific culture, often when predictions of the future are demanded a still large number of people resort to (variations of) prescientific ideas, e.g., horoscopes can be found in almost all newspapers. But this is a subject for anthropology and psychology, which are not in the scope of this book.

Given the importance that predictions had in our history, and have in our present and in science, it is worth briefly discussing "prediction" as a subject. We will do this with some simple examples, namely, by examining the following (non-exhaustive) list of desired predictions:

(a) What will be the position of the Moon tomorrow at midnight?
(b) Will be raining on the top of Campidoglio Hill, Rome, tomorrow at half past 12? or in 2/3 weeks at the same time?
(c) What will be the (much desired) winner numbers in the lotto game next week?
(d) What will be the value of the option of the gold in the London Stock Exchange tomorrow?

Clearly the answers to the above questions will be very different reflecting the differences of the underlying character of the phenomena of interest. The first question is rather easy to answer: one can give the proper answer even for each day of the next centuries, it is enough to look at some astronomical tables, nowadays we can find them even on the web. Experts working in meteorology are able to give an answer to the first part of the second question with some confidence, but surely will prefer to refrain from answering the second part. For the lotto game, we have to accept, unfortunately, the impossibility to say anything. There is not an unanimous consensus about the possibility to answer the last question. Some scholars believe that finance is ruled by the rational behavior of the economic agents, and therefore predictions, at least for a certain time interval, are possible. Conversely, others claim that the Stock Exchange is basically ruled by irrational, and sometimes also unfair behaviors, and therefore finance cannot be considered a science with the same status of physics or chemistry. A brief discussion will help to understand the reasons of the difference in the possibility to perform a prediction in the various contexts represented by the above questions.

For sure the simplest case is when there exists a deterministic rule (the equations of motion) governing the phenomenon we want to predict. In such cases, knowing (with infinite precision) the present state of the system, namely, the set of the variables $X(0) = (x_1(0), \ldots, x_N(0))$[1] describing the system, at the initial time $t = 0$, the future state $X(t)$ at any time $t > 0$ can be known, at least in principle, by solving the equation of motion (see the entry on Determinism). However, a moment of reflection would reveal that this idea is a bit naive even in a deterministic context as it requires: first, to know which are all the relevant variables that describe a given system, which is in general not obvious, second, to know their values with infinite precision, and when "infinite" appears in a practical context such that of performing a prediction some suspect should arise. Finally, it requires to know (with infinite precision) the evolution rule and to be able to propagate (with infinite precision and no matter the number of variables to be considered) the initial state with that rule up to the time at which we would like to know the state of the system. In the last requirements, there are so many infinites that one can start suspecting that even in the (simplest) case of a deterministic systems predictions are impossible. Of course, another moment of reflection should convince us that as (limited) human beings, on

---

[1] Where N can also be huge or even infinite, as when considering the velocity of the wind in the entire atmosphere.

a practical ground, a prediction will be possible only if replacing all the infinites of the above sentences with some finite precision and, in addition, approximating the variables and their evolution laws the future state can be forecast with a limited but satisfactory (according to some criterion) precision. To make concrete the above considerations, it is useful to reconsider the questions a) and b) of the above list.

In the case of the motion of the Moon, very accurate predictions were possible already in ancient times. For instance, the prediction of the eclipses was very precise already at the time of the Babylonian astronomers. They did not know that the Moon motion is ruled by Newton equation of gravitation, but they were very good and precise in observations and, fortunately for them, the orbit is rather regular, basically (almost) periodic, so the prediction is easy. Now we do not rely (only) on observations to predict the motion of the Moon or of other planets, we know the Newton equation and given the observation of the present state (reciprocal position of the celestial bodies) and of the parameters of the problem (the mass of the bodies, etc.) we can compute their future positions.[2] At this point we may object that we do not know with infinite precision the state of the system nor the parameters (we cannot weigh the Moon or Jupiter!), so why can we be so accurate in such predictions? The reason is that, as already said, the motion of bodies in the Solar System is quite regular and (more precisely) errors on the state and parameters of the system do not lead to the error on the future state of the Solar System to grow arbitrarily large in time, unless considering very long times, i.e., millions of years. As discussed in the entry Chaos, this is not always the case. In generic (non-linear) deterministic systems even a very small error on the initial condition (the present state) can grow in time exponentially with a rate given by the Lyapunov exponent, $\lambda > 0$. In other terms, if the error on the state of the system is $\delta(0)$ at time $t = 0$ it will become $\delta(t) = \delta(0) \exp(\lambda t)$ at time t. If Δ is the maximum error you are going to accept, the predictability horizon T_p will be determined by the request $\delta(T_p) = \Delta$, and thus using the above formula for $\delta(t)$ we have $T_p = 1/\lambda \ln(\Delta/\delta(0))$. The exponential rule is thus pitiless, for instance, if $\Delta = 1$ to make T_p four times longer you need to decrease $\delta(0)$ by a factor $e^4 \approx 54.6$. Therefore, the main player in determining T_p is the inverse of the Lyapunov exponent. For instance, the Solar System is also a non-linear chaotic system but, likely for us, its Lyapunov exponent is very small, therefore the prediction time, given by $1/\lambda$, is extremely large, roughly in the order of 10^7 years.

The problem of weather forecasting is conceptually similar to the prediction of astronomical phenomena: the evolution laws are deterministic and can be found by solving a set of partial differential equations. However, in the dynamics of the atmosphere one has to face two additional severe difficulties. The first one is that the Lyapunov exponent is much larger than in the Solar System: its inverse can be less than a second in generic conditions. This should make you ring the bell, we said before that the predictability time is essentially dominated by the inverse of the Lyapunov exponent, thus seconds or less! So how can a meteorologist even imagine to forecast

---

[2]We underline that the deterministic rule with link $\mathbf{X}(0)$ to $\mathbf{X}(t)$ in the most cases is not explicit, often the existence of such a rule follows from some general mathematical properties of differential equations and can be found, with (relatively) facility with numerical computations.

the weather over the Campidoglio Hill in one day or a week? The reason is connected with the second difficulty, which is the multiscale character of the atmosphere, i.e., the presence of many different time and spatial scales: the atmosphere is gaseous and is a fluid, where turbulence with its multiscale character (as seen in the entry Turbulence) is usually at play. Essentially the Lyapunov exponent is roughly given by the inverse of the fastest time scale in the problem, which relates to the smallest scales of the system (order of milliliters or even smaller). But, we are not interested in predicting such scales, we are usually interested in much larger scales: the weather over a town, or in a region. Such larger scales are characterized by slower time scales (a few hours, or a few days), and so predictions are difficult but conceptually possible over these longer scales. Thus, ironically, the multiscale character of the problem makes it more difficult in terms of understanding and modeling, and it allows for the possibility of making longer predictions. Because of the above technical difficulties, even for predictions on a limited time interval (say 1 week), it is necessary to master such a complex system with a clever use of mathematics and numerical algorithms, see, e.g., the entries Richardson, and Laws, levels of descriptions and models.

We now turn to the million euros question, namely, the winning numbers in the lotto game, which is plagued by the persistence of wrong dangerous superstitions about claimed hidden regularities of the sequence of the numbers which appeared in the lotto drawings (e.g., the infamous "theory" to play the most overdue numbers). Ironically, as our brain functionality of predictive processing helps us in dealing with many difficult tasks, again it seems that our brain blatantly fails in handling (at an intuitive level) probabilistic concepts. The error many commit in relying on overdue numbers in lotto is related to the so-called gambler's fallacy, namely, the misbelief that if some event has occurred more frequently than normal in the past it is less probable to occur in the future (or vice versa). Again this is an interesting subject for cognitive psychology but it is out of the scope of this book. Here we limit ourselves to observe that the dream to build an optimal strategy for the lotto game is hopeless. The situation is quite trivial: due to the independence of drawings the search of any rational strategy is meaningless, simply it is not possible to predict the winner numbers.

We can now examine the last question we posed about gold options in the stock exchange, and more in general about predicting the financial market. Just as working hypothesis, let us assume that finance is not ruled only by irrational and/or criminal behaviors. Even with such an assumption we cannot invoke a deterministic approach in terms of differential equations as for celestial mechanics and meteorology. It is more sensible to try to determine the price of an option in a probabilistic context, typically using stochastic models which use both the knowledge of the past and theoretical arguments. In a probabilistic approach one has to abandon the deterministic point of view in favor of a less demanding statistical study. Instead of asking "what will the state of the system be at time t?" we should ask "what is the probability that at time t the state of the system will belong to a specified subset?" Perhaps the most successful example of the probabilistic approach is the statistical mechanics (see the entries Statistical Mechanics and Universality), where it is possible, using the gigantic number of particles in a macroscopic body, to use the limit theorems of probability.

It is interesting to note that probabilistic approaches can and actually are used also in deterministic contexts. Since a few years when reading a weather prediction one should have noticed the presence of a probability over the forecast, e.g., tomorrow at ten will rain over Madrid with probability 60%. How it comes that such probability can be assigned? This is possible because researchers in meteorology have learned that, given the chaoticity, the uncertainty on the knowledge of the initial weather conditions and the need to use approximate models, the best strategy to cope with weather forecasts is to run their models with perturbations (i.e., tiny variations of parameters and state variables) to better represent the uncertainties so to generate an ensemble of forecasts over which probabilities can be assigned.

Let us now briefly discuss the problem of model building for the prediction of a complex phenomenon as for biological or medical phenomena, the task is decidedly more difficult, not to mention social sciences. For instance, can we approach the problem of predicting the evolution of the abundance of a group of species or the evolution of an epidemic (unfortunately, the latter being a theme of current concern)? In these examples, clearly there is nothing like Newton's equations for the mechanics or Maxwell's laws for electromagnetism. In the case of epidemics, we have nowadays (sadly) learned many technical terms such as the basic reproduction number R_0 and other quantities, but it is often unsaid or misexplained that these numbers and quantities are intrinsically probabilistic (so that their estimate is subject to large uncertainty) and that the predictions one can obtain are strongly dependent on many details of the specific (stochastic) model used to describe the evolution of the contagion. Epidemics are indeed stochastic phenomena so that only ensemble and probabilistic predictions make really sense, with all the uncertainties intrinsic in the choice of the model used. In the former case, many may have encountered the term Lotka-Volterra model, which can be either deterministic or stochastic and which describes dynamics such as predator-prey or even more general communities of species in interaction. However, models of this kind have a different status from the Newton equations. In such cases, constructing a model can only be derived from some often profound intuition, sometimes suggested by analogies or empirical observations. We do not enter the details of this important and difficult topic: in the entries Volterra, Richardson and von Neumann, we briefly discuss some aspects of the approaches used to build models in ecological systems and for weather predictions.

In the last decades, we have assisted to an ever-increasing availability of huge amounts of data as well as of sophisticated methods for their retrieval and computational power available for their analysis. The existence of such new tools led to a persistent emphasis on the (claimed) birth of a new fourth paradigm in Science (beyond the traditional ones based on experiments, theory, and numerical computations) based on data only.

While we cannot deny the importance and usefulness of analyzing data, and also the successful[3] stories of using data mining for predicting (or influencing?) social

---

[3] Avoiding any moral consideration.

behavior. As briefly discussed in the entry Big Data, there are two main severe difficulties when approaching predictions based on data only, for example, for weather forecasting. The first one relates to the size of the file necessary to use just data increases exponentially with the dimension of the system. As a consequence, a purely inductive method can work just up to dimension 5 or perhaps 6, but is deemed to fail with larger dimensions—this is often dubbed *the curse of dimensionality*. The second one concerns the difficulty to individuate the proper variables relevant to the phenomena under investigation. For instance, in the case of the prediction of the Moon's position, one has to consider also the variables related to the Sun and the Earth, in a similar way for the prediction of the rain over Rome it is necessary to take into account also the air temperature, humidity, velocity, pressure, and so on.[4] The above conclusions are not mysterious: they follow from the basic structure of Newton's mechanics and dynamics of fluids.

---

[4]As for this second difficulty, some help can come with the use of machine learning techniques, but discussing such a hot topic is out of the scope of this book.

Chapter 24
Probability

It is difficult to overestimate the importance of probability theory given the pervasiveness that chance and uncertainty have in all aspects of human activities, from the most practical to the most theoretical one. The great scientist Pierre-Simon de Laplace, in his book *Théorie Analytique des Probabilitiés* (1812), wrote *"It is remarkable that a science, which began with consideration of games of chance, have become the most important object of human knowledge"*. Coming to more recent times, the mathematician and Fields Medalist David Mumford in an interview stated *"The straitjacket in which logic forces the world has been useful at the beginning to allow mathematics to get the first results and to build mathematics as an intellectual enterprise [...] Over time, however [...] we are realizing that the world is better understood, in general, with probability and statistics"*.

A naive and intuitive idea of probability is innate in humans since this increased the chances to survive in the dangerous and complex prehistoric world, to develop the ability to deal with all those events, good or dangerous, occurring with a certain regularity (often predators hunt at night; at certain time of the year, frequently after rain, edible mushrooms grow, etc.). The words "often", "frequently", "usually", "sometimes", "rarely", etc. are frequency adverbs that qualitatively express our beliefs that an uncertain event takes place. In other terms, these adverbs indicate what we think to be, based upon experience, the probability for an event to occur.

Everyday we all make decisions based on subjective evaluations of probability of uncertain events—should I take my umbrella? buy an expensive gift? change my work? etc.—taking into account the certain costs and the possible effects of our actions. On the other hand, our intuitive perception of odds is subject to multiple cognitive biases that literally impede us from having a correct judgement of the probabilities involved in a given event.[1] Therefore, a basic knowledge of the principles of probability theory is important for everyday life considering that, when we

---

[1]Concerning this point, already Laplace in 1814 in his *Essai philosophique sur les probabilités* dedicates a whole paragraph to the illusions in the estimation of probabilities, and in recent times there have been many studies on cognitive biases that do not allow a correct instinctive evaluation of probabilities (see, e.g., the very interesting *Thinking, Fast and Slow* (2011) by Daniel Kahneman).

© Springer Nature Switzerland AG 2021
M. Cencini et al., *A Random Walk in Physics*,
https://doi.org/10.1007/978-3-030-72531-0_24

consult weather forecast, are dealing with medical diagnosis, weigh up the premium of an insurance contract, bet on the results of a sporting event, or buy tickets for a lottery, we always deal with probabilities.

Moreover, probability theory is fundamental for basically all scientific disciplines from quantum mechanics to genetics and medicine, from statistical physics to economics. In recent times, various scientists think of random variables as the founding elements of both logic and mathematics, again in the words of David Mumford *"probability theory and statistical inference now emerge as better foundations for scientific models, [...] and as essential ingredients of theoretical mathematics, even the foundations of mathematics itself"*.

One of the most ancient examples of evaluation of the probability of an uncertain event is the case of Bottomry, a maritime insurance contract. Such an insurance were stipulated by Babylon merchants (and also in use in the Hindus culture and in ancient Greece), by contract the owner of a ship borrows money for equipping or for repairing the vessel with the provision that if the shipment was lost in a specified voyage or period, by any of the perils enumerated in the contract, the loan did not have to be repaid. Obviously, if the voyage is successfully completed, then the owner of the ship will repay the debt plus an amount agreed in the contract. One can easily realize that the contract is a simple bet in favor of the success of the trade, and the lender must have the most accurate intuition of what the chances of success are.

Of course, probability issues also show up in games of chance that are as old as the human culture. One of the most ancient board games is the "hounds and jackals" that was played in the Ancient Egypt as early as 3500 B.C. In this game, the players have to move gaming pieces (small sticks with jackal or dog heads) according to specific rules by throwing some astragali, which are ancient four-sided dice made from a bone in the ankle, usually of a sheep.

From a more conceptual point of view, the origin of the debate about chance can be traced back to Leucippus, who affirmed that the atoms follow casual and unpredictable movement (that later became the "clinamen" of Lucretius) and Aristotle, who identified different kind of reasons for events to happen, that is, spontaneity, necessity and chance. The philosophical discussion about chance is also related to determinism, and we refer the reader to the dedicated entry.

The path toward the modern conception of probability theory is very interesting but not appropriate for this entry, and therefore we address the reader to one of the many good books dealing with the history of probability (e.g., see *History of Probability and Statistics and Their Applications before 1750* (1990) by Anders Hald). In our presentation, we jump directly to the sixteenth century when the Italian mathematician (and also physician, physicist, astrologer, gambler, ...) Gerolamo Cardano in his book *Liber de Ludo Alea* gave the first definition of fair dice, equiprobability and probability as the ratio of favorable outcomes to possible outcomes (when the possible outcomes can reasonably be considered as equiprobable) and his book can surely be considered the first text containing exclusively probabilistic problems. Cardano anticipated many problems and rules of probability that later gained a central role, but the lacking of a symbolic mathematical language prevented him from making a more incisive contribution. Also the father of modern science, Galileo Galilei, wrote down a short essay on playing dice, *Sopra le scoperte dei dadi*, in which he computed the probability of the possible sums of three dice. The problem had already been solved by Cardano, but Galilei, unlike Cardano, expressed the solution with great clarity, without errors, and with a unitary vision of the various aspects involved. In the seventeenth century, a French writer (and overall gambler) Antoine Gombaud, Chevalier de Méré, proposed the problem of points (i.e., the problem of the division of the stakes in a chance game when it is interrupted before the conclusion) to Blaise Pascal, another great mathematician (as well as physicist, writer, and theologist[2]). Pascal became quite interested in the problem of points and involved Pierre de Fermat in solving it. Fermat was a French lawyer without specific training in mathematics and started mathematical studies in his late 30s, anyway the results he achieved in his life have been amazing. The correspondence between Pascal and Fermat regarding the solution of the problem of points laid at the foundation of the mathematical theory of probability. Actually, the first text that can properly be considered a mathematical text on probabilities, *De Ratiociniis in Ludo Alea* (1657) of the great physicist Christian Huygens, was mainly based upon the corresponance between Pascal and Fermat. In his book, Huygens introduced the concept of expected value (a key concept in modern probability theory) specifically invented to solve the problem of points. The expected value of a random quantity is nothing but its average value over a very large number of trials. In particular, for the problem of points, the random quantity is the winnings for each player in case the game ends, and the expected value is used to quantify the expected winnings obtained by each player by completing the game. Using the expected values of the winnings for each player it is possible to fairly divide the stakes. For half a century, the booklet of Huygens was the only text treating probability with mathematical rigor and can be considered the real beginning of probability theory as a mathematical subject.

Our last note about the history of probability concerns Jacob Bernoulli and his book *Ars Conjectandi* published in 1713, 8 years after his death. Bernoulli was a member of a family of prolific mathematicians who sowed many important contributions in various research fields. His contribution to the initial development of the probability theory was seminal, since in his book, for the first time, the definition of probability as the ratio of favorable outcomes to possible outcomes was criticized since it can be applied only to a very few cases, almost exclusively related to gambling. Bernoulli conjectured that the geometric regularity of dice, coins, etc. has been chosen specifically to make games fair by ensuring equiprobability in the possible outcomes. But it is not possible to apply the concept of equiprobability to physical or social problems. The proposal of Bernoulli is to define a probability value of a given event in retrospect, as a result of repeated experiments based on the occurred frequency of that event. This brilliant scientist rigorously justified his proposal by means of a fundamental theorem, the law of large numbers, which states that as the

---

[2]It is not possible not to mention here the famous Pascal's Wager that links together theology and probability affirming that a rational person should believe in God:
We know neither the existence nor the nature of God... Let us weigh the gain and the loss in wagering that God is. Let us estimate these two chances. If you gain, you gain all; if you lose, you lose nothing. Wager, then, without hesitation that He is.

number of experiments increases, the frequency of a given event is indistinguishable from its probability whenever it can be explicitly calculated.

In this short excursus on the origin of the theory of probability, some important elements have been presented, but before discussing other mathematical and applicative aspects of the theory let us point out that there is an intrinsic dichotomy in the theory of probability. On the one hand, we have the mathematical aspects of the theory, i.e., axioms, properties, operations, theorems, etc.; on the other hand, we have the concept of probability itself, i.e., what is the value of confidence one can give to a certain event. We shall start by presenting the mathematical foundation of the theory that has to be consistent independently of the value of probability of the single event, for example, it must work well both for a fair and rigged dice, and later discuss about the definition of probability.

Though Laplace's book *Théorie analytique des probabilités* presented for the first time in a rigorous way the mathematical entities of the probability theory and their main properties, in the following, we privilege the exposition that can be found in the book of Andrej Nikolaevič Kolmogorov, *Foundations of the Theory of Probability* (1933), which maintains, almost one century after its publication, its freshness and depth for such a delicate subject. Herebelow we just give an informal introduction to the modern probability theory and we address to Kolmogorov's book the interested reader, see also the entry Kolmogorov to fully appreciate the stature of this important mathematician.

A modern mathematical theory starts by introducing the elements of the theory and then the operations allowed on those elements. The elements of the probability theory can be found in the space of the events, that is, a collection of sets representing all the possible outcomes of an experiment. For instance, in the lottery draws, the set of possible elements can be all the potential couples or triplets of different numbers between 1 and 90, i.e., the space of the events contains all the possible outcome of a lottery draw. The operation on the elements is those permitted by set theory, that is, union, intersection, etc. and the result of each possible operation is still an element of the space of the event. For example, in a roll of a die there are the elementary events 1, 2, 3, 4, 5, 6 but also the event "even" stays in the space of the event and it can be obtained as the union of the events 2, 4, and 6. Then, after the introduction of the space of the events, it is necessary to introduce the probability on such space. After Kolmogorov, such probability is nothing but a function (usually called $P(\cdot)$) that associates at each event a real number between zero and one. Zero is assigned to an impossible event (an event that cannot occur, e.g., 7 in the rolling of a cubic die). One is assigned to an event occurring with certainty (for instance, the event "any number smaller than 7" in the die example). Finally, one of the most important concepts of probability theory is that the value of the probability of the disjoint union of events (i.e., two events without any intersection) is equal to the sum of the probabilities of the events: $P(A \cup B) = P(A) + P(B)$ if $A \cap B = \emptyset$. For instance, in a roll of a die we define two disjoint events $A = \{1, 2\}$ and $B = \{6\}$, we must have $P(A \cup B) = P(A) + P(B)$. This last example can be easily verified using the additional notion of equiprobability, where $P(A) = 2/6 = 1/3$, $P(B) = 1/6$, and $P(A \cup B) = P(1, 2, 6) = 3/6 = 1/2$. Let us point out that in the approach of Kolmogorov there is no notion of equiprobability and the probability value is introduced by assigning the function P which must verify the aforementioned properties. Using these very few elements, Kolmogorov managed to construct the whole probability theory, achieving all the important results previously obtained.

In the Kolmogorov approach, there is no need to assign numerical values to the probability function in order to obtain all the important results and theorems of the theory. However, without such values it is impossible to calculate the odds of any event. Unfortunately, the definition of the probability of a given event can be a thorny problem since different events can actually have very different nature. Strictly speaking the lack of certainty in the outcome of an event is called uncertainty, but there are different types of uncertainty which may require different definitions of probability. For example, we have seen that for simple events (as "head" in a fair coin toss) Cardano's definition of probability, as the ratio of favorable to possible outcomes, may be fine. For slightly more complicated cases (as the event "eleven" in throwing a couple of rigged dice) Bernoulli's definition of probability, as the frequency of occurrence of a certain event in a sequence of repeated experiment, is more appropriate. But what about situations in which it is impossible to carry out controlled repeated experiments: how can we assign the probability of the event "win" for a given sport match, to the event "survival" in a car crash? or in much more complicated cases, such as the probability of a marriage to last or the event "evidence for extraterrestrial life" in the SETI (Search for Extra-Terrestrial Intelligence) program. In these instances, it does not seem easy at all to find a convincing way to assign the probability for a given event. A reasonable way to give a value of confidence to the occurrence of such complex events is to ask to a panel of experts that, using all available means, e.g., experiments, numerical simulations, historical data of similar events happened in the past, manage to express by mutual agreement a value (or range of values) of probability for the event.

The debate about the different interpretations of probability concerns genuine philosophical questions and many scientists and philosophers have been interested in this problem. There is no room here to enter the details of this deep issue. We shall limit ourselves to a short discussion and few comments. The main interpretations of probabilities are (i) the *classical* interpretation, that is, basically Cardano's one, is appropriate when it is trusted that the elementary events are equiprobable; (ii) the *frequentist* interpretation, introduced by Bernoulli, can be used when repeated experiments can be performed under the same conditions, (iii) the *subjective* interpretation, that is, the one that uses all the knowledge available (by any means) about the events and considers acceptable, in defining probability, also personal beliefs. Obviously, in order to make personal opinions compatible to a scientific discipline an additional principle is needed, namely, the *coherence* principle (introduced by Bruno De Finetti,[3] the founder of the subjective interpretation of probability) that

---

[3]The most famous sentence of De Finetti is *"Probability does not exist"* which must be interpreted as the probability is no way objective since, in his opinion, it is not an intrinsic characteristic of any real experiment (there exists no perfect coins, perfect dice, or experiments exactly conducted under the same conditions).

establishes that probability values for the events of a certain experiment is coherent if, considering all available information, a wager on such experiment cannot be unfair, i.e., it cannot result in a sure (or at least an average) gain. Unsurprisingly, the subjective interpretation of probability is very much pursued in social sciences such as economics.

It is worth noting that when all the interpretations of probability are applicable they return exactly the same values, and that, most importantly, the role of information and knowledge plays an absolutely central role in defining probability. For example, in conditions of complete ignorance, the probability of the event "head" in a coin toss must be fixed to 1/2, but only an insane man could continue to assign 1/2 to that probability after one hundred tosses resulted in one hundred heads. Moreover, not knowing anything about the fairness of a die, but knowing that the person who throw the die is the same of the fake coin of the previous example, we should be very careful in considering the elementary events as equiprobable. The subjective interpretation of probability always deals with conditional probabilities on the current state of our knowledge, while for the classical or frequentist interpretation, conditioning is usually not explicit because knowledge is supposed to be general and shared. Which is the most correct interpretation of probability is an ill-posed problem. Kolmogorov affirmed that it is meaningless to establish a priori whether or not it is possible to assign a probability to an event, this must be verified on a case-by-case basis. The best strategy is to use a pragmatic approach adopting the most suited method depending on the problem under examination.

In physics, the frequentistic interpretation of probability is typically adopted. Some branches of physics, as quantum physics, are based on probability theory, while, in other cases, probability theory is necessary to cope with some intrinsic limitation. For instance, in statistical mechanics, the probabilistic approach is imposed owing to the huge number of degrees of freedom involved (see the entry Statistical Mechanics and Entropy), while in chaotic dynamical systems due to the unavoidable limited knowledge on the initial conditions (see the entry Chaos). Probability is also a key element in Information Theory (see the related entry).

We conclude this entry with two examples of applications of probability theory taken from everyday life. It is impressive that in 1795 in his lesson on probability at the École normale supérieure Pierre-Simon de Laplace starts with *"In this lesson, I will talk about probability theory, interesting theory by itself and by its many relationships with the most useful objects of society"*, and today we still fail to give the right importance to probability theory. For the everyday life, possessing a good probabilistic culture implies, among other things, not being fooled by some false illusions and fake news. One of the most impressive collective illusions of the whole humanity is the possibility to become rich by winning at lottery. Almost all countries have a national lottery that promises to make people rich. Actually, someone gets rich, indeed at least one person will necessarily win the lottery, but the right question is: what is the odds that it could be me? In more fitting probabilistic terms, is the bet I make when buying a lottery ticket fair? It is surely useful to define what is a fair bet: given a prize, K, and the probability to win it, p, if the price I have to pay to enter in the lottery is $M = pK$, then the bet is fair, which means that playing the lottery many times, the winnings average is as much as the price payed to play the lottery. Just to make a simple example, no one would pay more than one dollar to enter in a coin toss bet in which the payout is two dollars. Indeed, $M = 1$, $K = 2$, and $p = 1/2$ and the bet is fair. Unfortunately, lotteries are very unfair and thanks to the work of Cardano and Laplace we know how to quantify their unfairness degree, that is, to compute the difference between the stakes of the player and the dealer multiplied by the respective win probabilities. For instance, in the case of the bet in a single number at the roulette, the player could win his bet 35 times, while the dealer could win the bet of the player. Calling b the amount a player bets, one has

$$b \cdot 35 \cdot \left(\frac{1}{37}\right) - b \cdot \left(\frac{36}{37}\right) = -\frac{b}{37},$$

where $(1/37)$ is the winning probability for the player and $(36/37)$ is the winning probability for the dealer. Therefore, fixing the bet b to 1, the unfairness degree of the roulette for the bet on a single number is $-1/37 \simeq -0.027$. It is interesting to note that the same result is obtained for the bet on red or black, for which $b(18/37) - b(19/37) = -b/37$. The unfairness of the roulette is very low, while for typical lottery the unfairness degree is much higher, of the order of unity. Since all lotteries are unfair and the dealer on the long run always wins, it is not really smart to play such a game, but if one really likes to bet, then using probability theory at least it is possible to calculate which game is less unfair and choose that one.

An aspect of probability theory that is difficult to grasp without a solid probabilistic culture[4] is the concept of statistical independence. Two events are called independent if the outcome of one does not influence the outcome of the other. For independent events, the rule of the product of probabilities applies: $P(A \cap B) = P(A)P(B)$. The difficulty to grasp such a concept can be easily illustrated by two examples. The first is related to the perception that if a number in a given lottery does not show up for many weeks, then it seems that it is more likely that at the next draw the number will come out. So many people have ruined themselves to follow this false perception of dependency: in a lottery each new game is independent from the past history, and the probability that a given number will be extracted is always the same and equal to all the other numbers (if the lottery is not rigged, obviously). Another tasty example concerns the fallacy of social forecast. About this, the election of Donald Trump in 2016 will remain famous in history, since his chance of winning computed by various companies specialized in opinion polling were extremely low: 2 days before the election day, pollsters and statisticians gave Hillary Clinton odds of winning the election between 75 and 99%. Also one of the most important companies in Election Forecast, the FiveThirtyEight of Nate Silver (a statistician famous for his innovative methods that allowed him to foresee some complicated cases) gave to Hillary Clinton 71.4% chance of winning. In a few hours, as the first projections on voting began to arrive, the odds of Donald Trump's victory passed from less than 20 to 100%.

---

[4]Which is the only way to really overcome the widespread cognitive biases, such as the *Gambler fallacy*, see the entry Prediction for a short discussion.

This complete debacle has had many causes, several works have been written in this regard, but one of the most important reasons concerns the difficulty in considering the degree of correlation (i.e., dependency) within a population. All forecasters know that polls are not perfect: the sample can be chosen unevenly, sometimes people do not respond sincerely, the distribution of turnout among different demographic groups cannot be easily computed, but part of a forecaster's job is to estimate how likely—and how extensive—correlated component could be. In the case of the American elections of 2016, few pollsters considered the correlation among the various aspects (e.g., concerning the turnout, many white men with a low level of education have decided to vote while younger educated people have decided not to vote, unlike what happened in the past by looking at the historical voting pattern) but in any case they did not consider the correct (very high) degree of correlation.

Chapter 25
Richardson

Lewis Fry Richardson (1881–1953) had been one of the great scientists of the twentieth century, but he does not belong to the club of scientists who are known at popular level (as Einstein or Hawking). Moreover, even though his name is linked to several seminal ideas and methods in different fields, many physicists and mathematicians often do not know who he was. Because of his deep integrity, in particular, of his pacifism, he did not follow the mainstream of his time, and often this had an impact on his career. Few words about the life and the beliefs of this creative scientist can help to understand and appreciate his contributions to science. His originality was often mistaken for eccentricity, and some of the ideas and methods that he conceived would be rediscovered only decades later. LFR had the tendency to be a bad listener, typically distracted by his thoughts, and during his life some people never took his work seriously considering him a funny man. G.I. Taylor, a great expert in fluid dynamics, wrote that Richardson *"seldom thought on the same lines as his contemporaries and often was not understood by them"*.

The last of seven children in a thriving English Quaker family, in 1898 Richardson enrolled in Durham College of Science. Two years later was awarded a grant to study at King's College in Cambridge, where he received a diversified education studying chemistry, physics, mathematics, meteorology, as well as botany and psychology. Because of the wide range of his interests, at the beginning of his career, he was uncertain which way to take. Later, he resolved to work in several areas, similar to the great German scientist Helmholtz, who was a physician and then a physicist, but following a different order. In his words, he decided *"to spend the first half of my life under the strict discipline of physics, and afterwards to apply that training to researches on living things"*; according to his autobiographical notes, he kept secret his program.

After graduation for 10 years (1903–1913), FLR held a series of short positions in several research laboratories, working on different topics, e.g., approximated solution of differential equations, meteorology, metallurgy, and percolation. Such a beginning of career was not uncommon at that time. In 1913, LFR was appointed to a position in the Eskdalemuir Observatory (southern Scotland). This position was rather attractive for LFR who had considerable freedom to pursue his main interest: a

© Springer Nature Switzerland AG 2021
M. Cencini et al., *A Random Walk in Physics*,
https://doi.org/10.1007/978-3-030-72531-0_25

method to compute the weather forecasting. During World War I, the Observatory was involved in military projects. Due to his pacifism, LFR was rather uneasy with this kind of research, and resigned on May 1916. Richardson was a Quaker and a conscientious objector, nevertheless, he took part (unarmed) in war operations by joining the Friends' Ambulance Unit, on the French front, transporting wounded soldiers, often under shell fire. Apparently, he was a mediocre driver but an excellent mechanic, with a great skill to solve practical troubles. It should be considered that during the war many leading scientists gave their contribution to the military effort, e.g., in aerodynamics (G.I. Taylor in England and L. Prandtl in Germany), in ballistic (J.E. Littlewood in England, V. Volterra and M. Picone in Italy), and chemistry (F. Haber in Germany). LFR was one of the few who deliberately ceased to do scientific research financed by the government during the war.

After the war, Richardson rejoined the Meteorological Office. Winston Churchill, at that time Air Minister, persuaded the Government to incorporate the Office into the Air Ministry, which was controlling the Royal Air Force. Unsurprisingly, LFR resigned again in 1920. Many of his colleagues expressed their regret and he continued to have with them friendly scientific collaborations. From May 1920, LFR was a lecturer at Westminster Training College, teaching to students for the bachelor's degree. In 1926, he was elected fellow of the Royal Society. For some years, LFR deliberately suspended his activity in meteorology: he was worried his research in turbulent diffusion could be exploited for military purposes. In 1929, he moved again at the Technical College in Paisley, near Glasgow. Paisley was rather far from the main centers of scientific activity and Richardson had rather intense teaching duties (16 h a week). However, LFR continued his research. The great meteorologist V. Bjerknes, as well other eminent scientists, expressed surprise that a scientist of such a stature did not have a more prestigious and secure position to continue his researches.

In 1940, thanks to a small inheritance, LFR retired early to devote himself to his studies in mathematical psychology, in particular, he was interested in using mathematical modeling techniques and statistical analysis to quantitatively investigate questions relating to war and peace. The same year he was offered a professorship. Although he had always longed for such a position, he refused it because (influenced by his deep pacifism) he decided to devote himself to understand the basis of international conflicts by means of a mathematical approach. He died in his sleep on September 30, 1953.

During his (almost hidden) life LFR gave astonishing contributions to science, often anticipating in a visionary way many important ideas and methods which are now widely used, and often known even at popular level. A non-exhaustive list encompasses: the idea to apply mathematics (in particular, numerical methods) in weather predictions; non-Gaussian diffusion in turbulence, which is now called Richardson diffusion in his honor; the idea of self-similarity and energy cascade in turbulence (see the entry Turbulence); the introduction of the fractal dimension (see the entry Fractals); the anticipation of the idea of parallel computing; and the first attempt to use mathematics to model the psychology of war.

It was during World War I that Richardson conceived his great, visionary idea of using the fundamental equations of fluid dynamics and thermodynamics to determine the future state of the atmosphere, which led to modern weather forecasting. At the time, the procedure conceived by Richardson was nearly impossible to perform, due to the lack of suitable computing tools, as there was no electronic computers.

Nevertheless, he managed to outline the problem in the correct way and also to define smart numerical algorithms (still in use today) to integrate the partial differential equations describing atmosphere dynamics. The manuscript of the book *Weather Prediction by Numerical Process*, written when off duty while on the front line, was lost during the battle of Champagne in April 1917. But, fortunately for the development of meteorology, it was fortuitously found, months later, under a pile of coal. The book was published in 1922 and it was not a commercial success.

Before Richardson, the weather forecasting was based on some semi-empirical methods, whose basic idea can be found in the Bible: *"The thing that hath been, it is that which shall be; and that which is done is that which shall be done: and there is no new thing under the sun"* (Ecclesiastes, 1:9). In a nutshell, the practical protocol was as follows. One looks for an analog, i.e., a past state "close" to that of the present: if it can be found at January 25, 1923, then it makes sense to assume that tomorrow the system will be "close" to January 26, 1923. Under the reasonable assumption that the weather is ruled by a deterministic dynamics, at a first glance the above protocol seems to be quite sound. However, it is not at all obvious that an analog can be found. The problem of finding an analog is strictly linked to the Poincaré recurrence theorem: after a suitable time, a deterministic system with a bounded phase space[1] returns to a state near to its initial condition, which means that analogs surely exists. But the relevant question, for practical applications as weather forecast, is how long have we to go back to find it? The answer, basically understood by Boltzmann while debating with Zermelo about irreversibility, is given by Kac's lemma. In a few words, the minimum length of the time series necessary to find an analog increases exponentially with the number of variables relevant to the dynamics of the system under consideration. In the case of weather, the number of relevant variables is huge, making the method of analog hopeless. These technical aspects are discussed in the entries Irreversibility and Prediction. Richardson, independently of Boltzmann, understood that the method could not work; there is no particular reason why we should find a previous analog: *"[...] the Nautical Almanac, that marvel of accurate forecasting, is not based on the principle that astronomical history repeats itself in the aggregate. It would be safe to say that a particular disposition of stars, planets and satellites never occurs twice. Why then should we expect a present weather map to be exactly represented in a catalogue of past weather?"*

In his attempts to weather prediction, Richardson introduced the main ideas on which modern meteorology is based. In his book, he suggested the use of the equations regulating atmosphere evolution: given a certain initial condition (today's weather), we can determine the future state (e.g., tomorrow's weather) by numerically integrating the equations. Now such an approach looks quite obvious (being currently used

---

[1] Which roughly means that the state variables cannot run to arbitrary large values.

in weather forecasting): we know that the atmosphere evolves according to the equations of hydrodynamics (for the fields describing velocity, density, pressure, water percentage, and temperature) and the thermodynamics giving the relation (equation of state) among temperature, density, pressure, and so on. Therefore, from the present state of the atmosphere, solving a set of partial differential equations we have a weather forecast. Of course, nobody can manage such equations only with the aid of pen and paper, the only practical possibility is a numerical attempt.

In order to appreciate the great novelty of his approach, one has to recall that the methodology of LFR was completely out of the meteorological school of Bjerknes,[2] which was dominant at that time. The weather predictions were performed with empirical and graphical schemes based on discontinuities (fronts) which, of course, cannot be handled with equations. LFR attempted to use in practice the method he conceived. His effort was titanic: as initial conditions Richardson used a record of the weather conditions observed in Northern Europe at 4 A.M. on May 20, 1910 during an international balloon day. The numerical work was long, taxing, and wearisome, his unique possibility was computing by hand with the poor help of some rudimentary computing machine. During the war, in the spared time around his duties (at the Friends's Ambulance Unit), in the course of 2 years he worked for at least one thousand hours. The result, a 6-hour forecast, was quite disappointing, but Richardson, even after the poor results, was moderately optimistic, and he (correctly) believed that his approach was the proper one, so he did not give up on his project, concluding that *"perhaps some day in the dim future it will be possible to advance the computations faster than the weather advances. ... But that is a dream"*.

Actually he went even farther, in spite of the fact that, before the World War II, only human[3] "computers" were available, he had the vision of the modern parallel computers. Pursuing his dream of weather predictions he imagined that several (human) calculators could simultaneously work in a "Weather Forecasting Factory": *"Imagine a large hall like a theater ... A myriad computers are at work upon the weather of the part of the map where each sits, but each computer attends only to one equation or part of an equation. The work of each region is coordinated by an official of higher rank"*.

Richardson's key idea to forecast the weather was conceptually correct, but in order to put it in practice it was necessary to introduce several further ingredients that he could not possibly have known. Even using modern computers it is not as

---

[2]Bjerknes was a great meteorologist and while his school was using more or less methods based on the analogs, he was among the first to recognize the importance of equations, which he contributed to write. Moreover, he basically had the same dream of Richardson, in his words *"If only the calculation shall agree with the facts, the scientific victory will be won. Meteorology would then have become an exact science, a true physics of the atmosphere. When that point is reached, then the practical results will soon develop. It may require many years to bore a tunnel through a mountain. Many a labourer may not live to see the cut finished. Nevertheless this will not prevent later comers from riding through the tunnel at express-train speed"*. Well, Richardson can be surely considered the man who started to dig that tunnel with his hands and a pickaxe.

[3] As discussed in the entry Computers, Algorithm, and Simulations, typically women were employed as human computers.

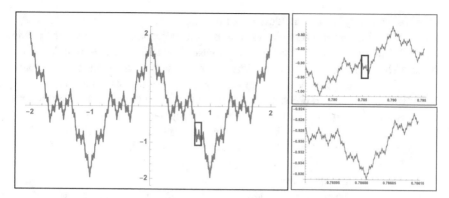

Fig. 25.1 The figure illustrates an instance of the Weierstrass function, obtained via the series $\sum_{n=0}^{\infty} a^n \cos(b^n \pi x)$, which is continuous and nowhere differentiable. The right panels display two consecutive zoom showing the self-similar character of the curves

straightforward as it may seem to numerically integrate the equations of hydrodynamics. One of the reasons is the presence of some numerical instabilities, which forces the use of very small integration steps. These instabilities are not just mathematical inconveniences, they have a clear physical origin: they are the consequence of atmospheric phenomena, such as sound waves and gravity waves, that have very short characteristic times. In order to actually overcome these shortcomings and obtain significant progresses, it is necessary to understand which aspects of the problem have to be taken into account and which can be disregarded.

The solution was found by von Neumann,[4] Charney and colleagues in the 1950s, within the Meteorological Project at the Institute for Advanced Study of Princeton (see also the entry von Neumann). The main conceptual progress was the building of effective equations that get rid of the fast variables. The introduction of a filtering procedure that eliminates processes not relevant for determining the weather has a clear practical advantage: the numerical instabilities are less severe, and so we can use a relatively large integration step, obtaining more efficient numerical computations. Moreover, the effective equations for the slow dynamics make possible to detect the most important features that, otherwise, would remain hidden in the original equations.

Already at the turn of the twentieth century, some mathematicians (Weierstrass, Peano, and Julia) investigated sophisticated constructions of curious mathematical monsters: very irregular curves which are continuous but non-differentiable. We can mention the von Koch curve and the Cantor set (see the entry Fractals) or the Weierstrass function (shown in Fig. 25.1). However, those mathematical objects were considered just as curious pathologies of no physical relevance. The widespread occurrence and importance of such behaviors in natural sciences were recognized

---

[4] von Neumann interest in the problem was not only scientific. He was pursuing weather modeling also having in mind a possible future ability to control the weather, so to use it as a weapon of war. A quite different point of view with respect to Richardson.

and popularized by Benoit Mandelbrot, who coined the term "fractal" and, correctly, acknowledged the credit to Richardson, to be considered the grandfather of fractals and self-similarity. LFR has been the first to ask *How long is the coast of Britain?*, this apparently silly question was explored in one of his papers found after his death, where he studied as a function of the resolution ℓ the length $L(\ell)$ of the coastlines of Great Britain, and other regions. Empirically, he found that rather than a convergence to a constant value, the length increases at increasing the resolution (i.e., decreasing ℓ) according to the law $L(\ell) \sim \ell^{-\alpha}$ with α increasing with the "wrinkledness" of the considered coast. After Mandelbrot, in modern terms, we know that $\alpha = D_f - 1$, where D_f is the fractal dimension. Nowadays, such a kind of wrinkledness has been recognized in many natural phenomena, where self-similarity is the dominating feature. Richardson seriously asked himself also another apparently strange question *Does wind have a speed?*, which led him, starting from just a few empirical data, to guess the self-similar structure of turbulence. He summarized his insights in the form of verses (inspired to a satirical one by Swift):

> *Big whirls have little whirls*
> *that feed on their velocity,*
> *and little whirls have lesser whirls*
> *and so on to viscosity,*
> *in the molecular sense.*

While studying the transport properties of turbulent flows, Richardson was among the first to realize the relevance of non-Gaussian processes. In 1905, Einstein first derived an equation for the diffusion of colloidal particles in a fluid at rest (see the entry Brownian Motion), which is characterized by a Gaussian statistics. More difficult is the problem of particles in turbulent flows, in particular, for their relative dynamics, that is, the evolution of the distance between two particles transported by a turbulent flow (see the entry Turbulence). This is not a mere academic problem, as it has important real-world applications. For instance, in the spreading of pollutants in the sea or in the atmosphere, it is important to understand how the patch gets larger in time. In turbulent flows, the diffusion equation derived by Einstein does not hold anymore, especially for particle pairs at separation within the scales at which the velocity is correlated. Richardson, with a deep insightful understanding of turbulence-related transport phenomena, proposed a diffusion equation in which the diffusion coefficient depends on the distance: the larger the distance the larger is the diffusion coefficient. All of this took place in the 1920s, about two decades before Kolmogorov proposed (actually influenced by Richardson's ideas) the first modern theory of turbulence. Remarkably, the solution of the diffusion equation proposed by LFR is not the usual Gaussian curve, in particular, one has that the large excursions are (relatively) frequent. Only at the beginning of twenty-first century rather accurate laboratory experiments and numerical simulations showed that the intuition of Richardson was correct.

Since Galileo nobody would have questioned the use of mathematics to describe the behavior of the physical world. Richardson somehow went farther and pioneered the use of mathematics to face problems outside the hard quantitative sciences. For instance, he explored the use of mathematics to model the conflict among states. It is difficult to say whether his approach is really able to catch the main aspects of these important problems; however, LFR with his enthusiasm and his genius had been a pioneer in the modern application of mathematics to social sciences, and anticipated von Neumann with his more sophisticated methods of game theory.

Chapter 26
Statistical Mechanics

Statistical mechanics[1] is an important part of theoretical physics. It aims at understanding physical systems made of many elementary constituents (that we can call particles for the sake of simplicity): a typical example is a substance made of many molecules such as a gas, a liquid, or a solid. Owing to the huge number of particles involved, it appears rather evident the practical impossibility to predict their precise behaviors (see also the entries Determinism and Prediction). Moreover, even if such a precise prediction was possible it would be useless in practice, as the properties of interest are usually associated to some average behavior of the particles. For instance, the electrical current results from the average flow of electrons in a metal and there is no need to determine the trajectory of each individual electron. If averages are the meaningful observables, it should not come as a surprise that probability and statistics come into play. Indeed, the distinguishing feature of statistical mechanics, with respect to other branches of theoretical physics, is its methodology which deeply grounds on probabilistic concepts and tools.

In the last decades, the number of applications of statistical physics has steadily grown. The methods of this discipline are now used to investigate not only traditional physical systems such as those made of molecules—that can be found in the non-living natural world—but also to biological systems, from proteins to insect swarms or bird flocks, to human activities, such as car traffic or economy, and also problems in between, such as epidemic spreading.

In order to understand the key difference between the traditional (Newtonian) mechanics and statistical mechanics, we invite the reader to imagine a Solar System made of a big sun and small orbiting planets, see Fig. 26.1 (next page). This system has been the fundamental test ground of Newton's laws of mechanics and of gravitation: the motion of planets and moons is fairly predictable and is characterized by

---

[1] "Statistical physics" is frequently used instead of "statistical mechanics". Some authors consider the former more general than the latter, which is usually associated to the probabilistic investigation of molecular systems (molecules form in fact "mechanical" systems). "Statistical physics" instead is used for all other systems (see the end of this entry). However, this is a minor point, and here we use the two terms as synonyms.

© Springer Nature Switzerland AG 2021
M. Cencini et al., *A Random Walk in Physics*,
https://doi.org/10.1007/978-3-030-72531-0_26

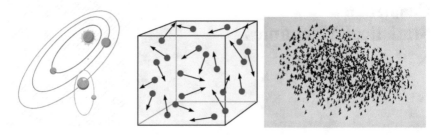

Fig. 26.1 Single components of a planetary system (planets and moon) are fairly predictable. On the contrary, a system of several interacting molecules is predictable only in its average properties. The second system is the historical subject of statistical mechanics. The third system is a flock of birds, which is a recent interest of statistical mechanics

superimposed periodicities, at least at a first level of approximation.[2] Things become more complicated if the number of planets increase. A system with large number of interacting planets is perhaps too strange for layman's imagination: the reader can replace the planetary system with that of a more familiar billiard, but now, at variance with the real case, the number of the balls is very large. Let us remove friction (it is an ideal experiment, done only in our head or within a computer), so that billiard balls—once in motion—never stop to run across the table. Several balls will frequently interact, by elastic collisions, among each others and with the billiard walls. Our imagination, as well as a computer simulation, tells that such a system with many balls behaves very differently from a planetary system: the billiard with many balls seems crazy, the planetary system is dull, boring, practically periodic. In a planetary system, it is possible to make calculations about the future movement of planets or moons in order to predict eclipses, tides, or plan spaceship missions. In the case of the billiard with many balls prediction seems difficult, on the other hand one has the feeling that we have some statistical regularities.[3] Basically, as a consequence of the large number of balls, although we are not able to perform a precise prediction, we have the emergence of statistical laws, for instance, we may predict the average frequency of impacts among the balls or the pressure on the walls (as in a gas, of which the many balls billiard is a macroscopic idealization). This latter information could be useful, e.g., to estimate when the wall will break.

---

[2] Actually, the Solar System is chaotic; however, its Lyapunov exponent is so small that, at least for some millions of years we can fairly assume a regular motion of the planets (see also the entry Prediction).

[3] The difficulty of making prediction on the motion of the individual constituents, the balls, in this case, are present whenever their number become too large, as in a gas. In the specific case of spherical billiard balls without friction and colliding elastically, we have in addition the presence of chaos (see the related entry) which would prevent long-term prediction also if the number of balls is relatively small. This means that also with a small number of balls a statistical description may be needed but for different reasons. When the statistical mechanics approach was introduced, see below, the existence of chaotic behaviors was not yet known.

The origin of statistical physics can be traced back to the work of Rudolf Clausius, Ludwig Boltzmann, and James Clerk Maxwell in the central and second parts of the nineteenth century (see the entries for Boltzmann and Maxwell). Before the work of the three founding fathers, the main known physical phenomena related to macroscopic bodies constituted of many interacting particles were explained in terms of fluid mechanics or thermodynamics. The first approach, older and rich of mathematics, described the statics and dynamics of fluids (gases and liquids) in terms of macroscopic quantities such as density, flow velocity, and pressure; the second, more recent and heuristic (see the entry Entropy), focused on temperature, heat, and transformations of energy. In both disciplines, the intuition or the hypothesis of a microscopic atomic world (see the entry Atoms) was (at least partially) present, but without strong consequences. An important example of a primitive form of statistical mechanics (more precisely of kinetic theory) can be found in Daniel Bernoulli's derivation of Boyle's law[4]: such derivation relies on the intuition of pressure as the average effect of impacts of many molecules on the container's walls. Bernoulli's theory (1738) was more or less forgotten for at least a century, until Rudolf Clausius built his own kinetic theory of gases (1857). Therefore, the fundamental concept of mean free path (i.e., the free space crossed by a molecule after a collision and before the next collision with other molecules) was introduced for the first time. Reading Clausius paper, Maxwell came to the idea of a probability distribution of the velocities of molecules in a gas, which is considered as the first systematic application of probabilistic concepts to physics.

A probability distribution (see the entry Probability) is a function that tells the probability of finding a certain outcome for a random event, for instance, the probability distribution of a (honest) six-faced die is a function that gives $1/6$ for every number between 1 and 6, which are the possible—identically probable—outcomes of throwing the die. In the late nineteenth century, statistics was a developed discipline with established roots in probability and important applications in social sciences: as its name says, it is a relevant tool for governments, politics, and economics, all aspects of life that involve large numbers. With the advent of nation states and modern public administrations, the properties of populations, births, deaths, diseases, commercial exchanges, migrations, etc. started to become interesting objects of study involving statistics. By measuring frequencies, it is possible to estimate the probability distributions for the outcome of future events and this is crucial, for instance, to plan investments or build infrastructures. Empiric evidence and mathematical theorems point out that certain probability distributions are more important and common than others, a chief example is the Gaussian[5] (or normal) distribution which is ubiquitous in nature as well as in several aspects of human life.

Based upon this knowledge, Maxwell in 1860 made the hypothesis for the probability distribution of the speed (the modulus of the velocity) of molecules in a

---

[4]An experimental law affirming that the pressure of a gas is inversely proportional to its volume, so that compressing a gas its pressure increases.

[5]As briefly discussed in the entry Universality, it derives from limiting laws which involve a large number of random variables.

dilute gas, a distribution which is strictly related to the Gaussian one (in fact, if the Cartesian components of molecules' velocities are independent Gaussian-distributed random variables, then the speed obeys the Maxwell distribution). Maxwell in the same papers foreshadowed the concept of equipartition of energy at equilibrium[6] and derived a non-intuitive expression for the viscosity of gases from dynamical reasoning. In the same years, Boltzmann, stimulated by Maxwell's results, developed an even more sophisticated and deep statistical theory of gases. This culminated in 1872 with the publication of its famous equation for the dynamical evolution of the velocity distribution of molecules in a gas. The equation is important and profound for several reasons. First, it generalizes to non-equilibrium states the idea of applying probability concepts to the mechanics of molecules. Second, it demonstrates that from any initial state, if left unperturbed, a gas will relax to an equilibrium state where the molecules' speeds are distributed according to the Maxwell distribution (which, in fact, is often referred to as the Maxwell-Boltzmann distribution). Third it anticipates (even if with some important differences) a class of equations, called "master equations", which are the backbone of the theory of stochastic processes, a mathematical framework which will be developed more than 50 years later. These equations have the common feature of expressing the variation, in time, of the probability that a variable takes a given value, as the sum of a "gain term", due to events that increase the probability of that value, and a "loss term" due to events that reduce such a probability. Boltzmann equation is not important only for theoretical physics, but it is useful in several technological applications, for instance, in semiconductor devices and in aerospace engineering.

The first steps of statistical mechanics were confined to the physics of gases because in such systems the interaction among molecules was weak, apart during impacts, and in many cases could even be ignored without losing generality or good agreement with experiments. In a sense, we could say that the development of statistical mechanics in the twentieth century has been mainly focused on relaxing the hypothesis of weak interaction. Facing the challenge of statistical predictions in the presence of relevant interactions was made possible thanks to two other fundamental ideas of Boltzmann. The first is the generalization of the Maxwell-Boltzmann distribution for molecular velocities to the distribution of energies of any statistical system. A system of interacting molecules has an energy which is the sum of all kinetic energies of the molecules and of all mutual interaction potential energies, plus possible external potentials. The expression representing the energy as a function of all positions and velocities of the molecules is called "Hamiltonian" (in honor of the mathematician William Rowan Hamilton who formulated a general theory of mechanics in the 1830s): Boltzmann gave convincing arguments that the probabil-

---

[6]A gas, or any other system, is in thermodynamic equilibrium if there is no macroscopic flow of matter or energy (in any of its forms, such as, e.g., heat) within it. Energy equipartition means that any form of energy in the system is equally distributed between its degrees of freedom, for instance, for a monoatomic gas the average kinetic energy is equally shared among the atoms, and among the motion components of each atom. In formulas, assuming all atoms to be of equal mass m, we have $\langle \frac{1}{2}m|\boldsymbol{v}^i|^2\rangle = \langle \frac{1}{2}m|\boldsymbol{v}^j|^2\rangle$ for any pair of atoms i and j, and moreover $\langle \frac{1}{2}m|\boldsymbol{v}_x^j|^2\rangle = \langle \frac{1}{2}m|\boldsymbol{v}_y^j|^2\rangle = \langle \frac{1}{2}m|\boldsymbol{v}_z^j|^2\rangle$, where x, y, and z are the Cartesian components of the velocity of atoms.

ity of finding the molecules of any substance—at equilibrium temperature T—with given positions and velocities (a "microstate") is proportional to $\exp(-\beta H)$ where H is the Hamiltonian function of all positions and velocities of the microstate, and β is a constant that later was identified to be $1/(k_B T)$ where k_B is the Boltzmann constant (see the entry Boltzmann). One can show that the Maxwell-Boltzmann distribution for the velocity of the particles holds in any system. The second and more general (i.e., not restricted to rarefied gases) contribution of Boltzmann is the expression of thermodynamic entropy, S, in terms of the number (or probability) of microstates, W, the so-called tombstone formula, $S = k_B \log W$, as it is engraved on his grave (see Entropy entry).

Boltzmann's ideas on the application of probability to mechanics were later developed by Josiah Willard Gibbs (who lived and worked in the United States between 1839 and 1903) few years later. It is fascinating to realize that Gibbs was the first American theoretical physicist to earn an international reputation: United States were at the beginning of their development as a technological world power, and theoretical physics was not their first interest (nevertheless, only 40 years after Gibbs' death, the US led the world to the theoretical and practical discovery of nuclear power). Gibbs coined the name "statistical mechanics" and invented the concept of statistical ensembles (actually this concept was basically present in the works of Boltzmann), providing examples of calculation of thermodynamic properties, free energy, etc. in terms of microstate distributions: the expression $\exp\left(-\frac{H}{k_B T}\right)$ for the probability of microstates is often named Boltzmann-Gibbs distribution.

Crucial contributions to statistical mechanics were given by Einstein (who apparently did not know Gibbs' contributions) in his early scientific works. Einstein was more interested to non-equilibrium aspects (such as diffusion and Brownian Motion) but crucially contributed to equilibrium theory too, for instance, with his renowned fluctuation's formula based on Boltzmann's entropy, see the entry Einstein.

There is another fundamental approach introduced by Boltzmann that is somehow antithetical to that of Gibbs statistical ensembles. While Gibbs imagined to have many copies of the same physical system and therefore interpreted a statistical average as the average over all such systems, Boltzmann preferred to view averages as the result of long experiments over a single system. This is the so-called Boltzmann's ergodic hypothesis. Boltzmann imagined that a physical system made of many molecules evolves very rapidly and that—even if its macroscopic properties seem stable— the microstates (the exact position and velocities of all molecules) change continuously by exploring all possible configurations. The ergodic hypothesis—stated by Boltzmann—is the idea that an isolated mechanical system (for instance, a group of atoms) explores uniformly all possible configurations.[7] To help our imagination, let us think again of balls on a billiard table (preferably without holes and with negligible friction). The reader can imagine to put all balls in motion: they will collide among each other and with the perimeter of the table, moving randomly everywhere.

---

[7]Briefly and avoiding to enter too technical matter, we would like to emphasize that ergodicity is not strictly necessary to ensure a "good" statistical mechanics. Indeed it has been shown that to get meaningful averages in statistical physics, hypothesis less restrictive than ergodicity is sufficient.

Because of the absence of friction, energy is never dissipated and balls never cease to move; in this case, one expects that each region of the table is sooner or later crossed by each ball. This is, very roughly speaking, a realization of an ergodic system. Ergodicity for billiard was mathematically proved by Yakov Sinai considering a square billiard with a single ball provided that a circular internal obstacle is placed in the center of the square.

Ergodicity became, later, a fundamental concept in probability theory, perhaps more technical and abstract than in the original idea and formulation by Boltzmann. In physics, it was explored by several scientists (see, for instance, the entry Kolmogorov). Another important scientist who explored this concept in physics and who also gave other seminal contributions to the foundations of statistical mechanics was Paul Ehrenfest (1880–1933), student of Boltzmann. Apart from the theory of adiabatic invariants (related not only to ergodicity but also to quantum mechanics), Ehrenfest made key contributions to the theory of phase transitions as well as to the connection between statistical mechanics and quantum physics. The theory of phase transitions is one of the most fascinating parts of statistical physics, still developed today. The term phase transition denotes a strong change of the properties of a substance occurring when some macroscopic parameter is varied by a tiny amount: the main example is melting or evaporation, for instance, from ice to liquid water and from liquid water to vapor. Such huge changes of state of a substance are obtained by varying the temperature, T, by a fraction of degree, and this makes a theoretical investigation quite puzzling. What could happen, in the theory, at a particular value of T? To answer this question, a crucial concept of statistical mechanics must be introduced, that is, the so-called thermodynamic limit: this is the mathematical procedure—applied to statistical formula—of increasing the number of molecules N and the volume V to infinity, in a way such that N/V remains constant. This makes sense because the number of molecules in macroscopic substances is astronomically large, of the order of 10^{20} or more. When this limit is taken, certain smooth behavior may become singular and sudden transitions appear in the calculations. Another scientist who gave fundamental contributions to the theory of phase transitions was Lev Landau (1908–1968) who was awarded the Nobel Prize in 1962 for his explanation of superfluidity (a phase of very cold liquid substances such as helium).

The full development of statistical mechanics, particularly in the theory of phase transitions, required the development of new important statistical concepts, leading this discipline to evolve into a much more comprehensive theory, embracing systems outside of its original scope. For instance, in several situations, the elementary constituents of a statistical system (atoms, molecules, or even electrons, which may leave their atoms and flow in a metal carrying electrical current) are strongly correlated and the probability tools to describe them become very complicate. Long-range correlations (i.e., the fact that a perturbation in a region of a system may have strong and immediate influence in a distant region of that system) appear in substances which are in the proximity of so-called "critical" phase transitions as, for instance, the transition of iron from a non-magnetized state to a magnetized one. In the presence of long-range correlations, the statistical description of a substance is greatly simplified and certain properties of the substance become "universal" (see entry

Universality) meaning that many microscopic details become irrelevant and there emerges identical behaviors, also at a quantitative level, for large classes of very different substances. The study of long-range correlations and critical phenomena is strictly tied to the development of the concept of self-similar phenomena and fractals. Such concepts found applications well outside the classical scope of statistical mechanics, i.e., beyond the mere study of substances made of molecules, and mark the key transformation of statistical mechanics into statistical physics and, according to many, into "complexity theory".

Complexity is a word with an elusive meaning, which has been often abused by media and commercial science popularization. In our opinion, it has its true foundation in statistical physics,[8] where it is a synonym of emergence of a collective behavior which is qualitatively different from the behavior of the single elementary constituents. As it happens for phase transitions, where from the behavior of a single water molecule, it is not possible to prefigure at all phenomena such as evaporation or freezing, several other systems in nature (and not only) may reveal behaviors that cannot be foreseen by considering a single component. The popular slogan of this paradigm is "more is different", which is the title of a famous paper by the physicist Phil Anderson published in 1972 on Science, a paper where the idea of reductionism is challenged by real examples taken from statistical physics (see the entry Laws, levels of descriptions, and models).

Nowadays, the sophisticated statistical mechanics tools, originally developed in physics while trying to understand the emergent behavior of inanimate molecular substances, are commonly applied and adapted to well different systems. Statistical regularities which resemble universal behaviors of critical phase transitions appear in disparate phenomena such as those occurring in finance, geophysics (earthquakes and avalanches), erosion, friction between solids, several kinds of natural or artificial noises, rains, etc. Statistical physics today has also a very broad interest into biology, for instance, in molecular biology where it aims at understanding the mechanics of inner cellular mechanisms (DNA replication and error correction, protein synthesis, transports of substances through the membranes). Biology is tackled by statistical physics also at larger scales: examples of emergent phenomena can be found in all living aggregates such as colonies of bacteria, cellular cultures, but also in collective behaviors of superior animals such as bird flocks, fish schools, or insect swarms. Crowds of humans (e.g., constrained in small spaces), social networks, or car traffic are also phenomena studied under the lenses of statistical physics, with several interesting analogies.

An important—technical but not only—aspect of this explosion of applications of statistical physics is the necessity of a generalization of this discipline, and its foundations, beyond thermal equilibrium, which is the particular setting where the theory has been mainly developed during the twentieth century. The need—together with the difficulties—of a non-equilibrium statistical mechanics was clear also to

---

[8]There are of course other good definitions of complexity in specific disciplines, for instance, Kolmogorov complexity or computational complexity in algorithmic information theory or the theory of dynamical systems, see the entry Complexity for a full account.

the founding fathers (both Boltzmann and Einstein devoted a lot of time to this problem). If thermal equilibrium is absent, for instance, when the air of a room is influenced by both a heating radiator and a frozen window, the Boltzmann-Gibbs formula for the probability of microstates is no more valid, so that the entropy formula and all the apparatus of statistical mechanics to calculate free energies becomes useless. This happens also in a flock of birds or a flow of pedestrians (see, e.g., the entry Mesoscale systems for further discussions). While many progresses have been done in the development of non-equilibrium statistical mechanics and some general principles have been established we are still far from the level of completeness that statistical mechanics has reached in the context of equilibrium phenomena.

Chapter 27
Turbulence

Fluid turbulence is an ubiquitous and commonly experienced phenomenon and refers to the disordered, seemingly random and unpredictable spatio-temporal evolution of a fluid. We can observe it in a cap of coffee with milk stirred with a spoon, in the complex patterns of a puff of cigarette smoke, watching the water falling from a fully open faucet or in satellite images showing the development of weather perturbations. We can also experience it through the violent trembling of the airplane when passing across patches of turbulent air (as typically announced by the pilot). Even if not fully realized, we pay for the effect of turbulence whenever we use a car or an airplane, as most of the fuel is consumed by the turbulent drag, hence the obsession for the vehicles' aerodynamics.

The term turbulence originated in the Middle Ages to denote the disordered and unruled motion of crowds. The first scientific use of the term dates back to Leonardo da Vinci who, in the *Codex Atlanticus*, wrote *"Doue la turbolenza dellacqua rigenera, doue la turbolenza dellacqua simantiene plugho, doue la turbolenza dellacqua siposa"*[1] to describe the swirling motions of water in rivers. In a later manuscript,[2] he wrote *"The small eddies are almost innumerable, and large things are only turned round by large eddies and not by small ones, small things revolve both in small eddies and large"*. These two sentences are often regarded by many scholars as an anticipation of the modern view of turbulence that we will discuss below.

Fluid turbulence brings about a multitude of problems and issues in both fundamental and applied research, with often a blurred border, and have been approached with different perspectives (which changed with time) by both engineers and physicists. Consequently, providing a definition of turbulence able to satisfy the large community of scientists interested to the turbulent problem(s) is not an easy task. This is exemplified by the following joke circulating among the experts in the field:

---

[1] "Where the turbulence of the water is generated, where the turbulence of the water is maintained for some time, and where the turbulence of the water is lost".

[2] Leonardo, Paris Manuscript F.

© Springer Nature Switzerland AG 2021
M. Cencini et al., *A Random Walk in Physics*,
https://doi.org/10.1007/978-3-030-72531-0_27

"Turbulence is like pornography. It is hard to define but if you see it, you recognize it immediately". These difficulties can be appreciated in the different ways the term *eddy* is used in the specialized literature. Sometimes it is used as a synonymous of vortical structure, or more in general of a coherent structure, i.e., some patch of fluid whose velocity field has some recognizable (by eye) and lasting (for some time) degree of organization. While for others it has a looser meaning of velocity fluctuation or excitation at a given scale. In the former case, the stress is on the "object" in the latter on the statistical features of the fluctuations. Moreover, in studying turbulence, two complementary points of view can be adopted. We can focus on the fluid velocity field, as an observer that looks from a bridge the turbulent river below—this is the Eulerian point of view. Otherwise, motivated by the efficient spreading and mixing of substances (e.g., the cigarette smoke) transported by turbulent flows, we can focus on the evolution of a fluid element—this is the Lagrangian point of view. In the following, we will briefly discuss both viewpoints, starting with the former.

Often, looking at physics from outside, it may seem that understanding some phenomenon just means to find the laws governing it. With turbulence this is not the case as we know the equation of fluid motion since the 1820s thanks to Louis Navier (1785–1836) and George Stokes (1819–1903). The Navier–Stokes (NS) equation,

$$\partial_t \boldsymbol{u} + \boldsymbol{u} \cdot \nabla \boldsymbol{u} = -\frac{1}{\rho} \nabla p + \nu \Delta \boldsymbol{u} + \boldsymbol{F} \,,$$

is nothing but Newton's second law applied to an infinitesimal element of fluid with density ρ and velocity $\boldsymbol{u}$. On the left-hand side of the NS equation, we recognize the acceleration, while on the right-hand side the forces (divided by the density) applied to the fluid element: first, the force due to adjacent fluid elements, namely, the gradient of the pressure p; second, the dissipation due to friction among fluid elements, ν being the kinematic viscosity; and finally, the external stirring force, without which the fluid would reach a rest state due to dissipation. The above equation must be supplemented by boundary and initial conditions and by an equation for the fluid density here omitted as we assume the fluid incompressible implying a constant density.

The first systematic experimental investigations of fluid motion in laboratory-controlled conditions started about 60 years after the Navier-Stokes equations were introduced with the work of Osborne Reynolds (1842–1912). He was studying fluids flowing in pipes (with circular section). Pipe flows appear to be regular, the technical term is laminar, up to a certain rate of flow influx, and well described by an exact solution of the NS equation due to J. L. M. Poiseuille and G. H. L. Hagen who, around 1840, predicted a parabolic profile for the velocity field with maximal velocity U (which can be computed knowing the flow rate) in the pipe center and zero velocity at the walls. Experimentally, the laminar state can be visualized injecting a tiny jet of dye in the pipe center, which remains as a horizontal filament far from the injection point. However, upon increasing the flow rate, and thus the velocity U, the parabolic solution ceases to be valid, the filament breaks up and invades the whole section, displaying a spatio-temporal disordered evolution, i.e., there is a transition to

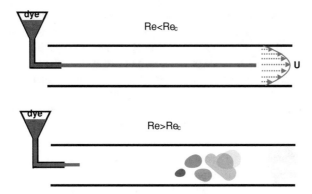

Fig. 27.1 Sketch of Reynolds experiments. (Top) Laminar flow for $Re < Re_c$ the dye injected in the pipe center remains well defined and unmixed till the end of the pipe, and the velocity field has a well defined parabolic profile. (Bottom) For $Re > Re_c$ the dye filament breaks up becoming well mixed far from the injection point, the velocity field (not shown) becomes very irregular

a turbulent state (see Fig. 27.1). While this was more or less known, the experiments of Reynolds showed (by varying the size L of the pipe section, the fluid viscosity ν, and the flow rate, i.e., U) that the transition occurs at particular value of a non-dimensional number $Re = UL/\nu$, now named Reynolds number.[3]

While well characterized experimentally, the nature of the transition to turbulence remained conceptually elusive up to 1971, when David Ruelle and Floris Takens wrote a famous paper titled *On the nature of turbulence* in which they understood the transition to turbulence as the appearance of chaos. At that time, chaos in nonlinear dynamical systems (see the entry Chaos) was starting to be a mature research subject, and it was known that upon varying some control parameter (i.e., Re for the Navier-Stokes equations) the system evolution from regular (e.g., steady fixed points or periodic motions) suddenly becomes chaotic. Ruelle and Takens also predicted possible routes of the transition to chaos that were, after a few years, confirmed experimentally in fluid experiments, demonstrating the chaotic nature of turbulence.

The chaotic and very high-dimensional (as clarified in the following) nature of turbulence implies that we can only hope to derive some statistical theory of it. Remarkably, this was realized much earlier than chaos entered the scientific playground. Given the blatant evidence of randomness and irregularity in the evolution of turbulent flows, already in the mid-1930s, Geoffrey Ingram Taylor (1886–1975) published a series of works in which he laid the seeds of modern statistical theories of turbulence. He started to look at a turbulent velocity field as a random field, which should be studied in terms of statistical quantities such as correlation functions between velocities at different points and times. In simple terms, these quantities provide information on how the degree of (statistical) coherence decays with time

---

[3]The importance of Re can be deduced as follows: if we make non-dimensional spatial coordinates with L, velocities with U, and time with L/U, one can obtain a non-dimensional version of the NS equation where the viscous term reads $Re^{-1}\Delta'\boldsymbol{u}'$, where $\prime$ denotes non-dimensional variables.

and with the distance.[4] Moreover, he introduced the important concept of isotropic and homogeneous turbulence which, strictly speaking, is something that in nature does not exist as there is always the statistical signature of walls or of the stirring mechanisms. Essentially, the idea is that if turbulence is developed (i.e., the Reynolds number is very large), far from the boundaries, turbulent fluctuations should have properties which do not depend on where they are measured (homogeneity) or on the direction (isotropy). So that, roughly, the intensity and properties of turbulent fluctuations should be the same in all directions with respect to any frame of reference.

A statistical theory of turbulence is feasible only if, at least for large Reynolds numbers, it is reasonable to assume that the statistical properties are independent of boundaries, forcing, and dissipation, i.e., if they are *universal* (see also the entry Universality). In this respect, the notion of isotropic and homogeneous turbulence fits the idea of universality. But why assuming such universality is reasonable? To appreciate this point we need the notion of energy cascade (sketched in the Fig. 27.2). When we excite a fluid we typically do it at quite large scales, e.g., the size of a car or the spoon in a cap of coffee. As already Leonardo recognized, a turbulent flow is characterized by eddies of all sizes and actually the dissipative term in the NS equation is large only at small scales, meaning that only the smaller eddies contribute to dissipation. Thus, there should be a mechanism bringing energy from the large (injection) to the small (dissipation) scale. The only term which can serve at this scope is the non-linear one, indeed one can show that it cannot change the energy but only redistribute it. Since this is an intrinsic process unrelated to forcing and dissipation, it is reasonable to assume that in this process of energy transfer from large to small scales details on the forcing and boundary conditions are forgotten leading to some form of universality. This idea of energy cascade was around since 1926 owing to Lewis Fry Richardson who expressed it in verses (see the entry Richardson) inspired to some verses of Jonathan Swift's poem *On Poetry: A Rapsody* (1733): *"So, nat'ralists observe, a flea/Hath smaller fleas than on him prey;/And these have smaller yet to bite 'em,/ And so proceed ad infinitum./ Thus every poet, in his kind,/ Is bit by him that comes behing"*.

With reference to the sketch of the energy cascade, see Fig. 27.2, energy is injected at large scales, say at an average rate ϵ_i, generating large eddies which give birth to smaller eddies and so on up to scales small enough that the viscous term becomes effective and dissipates the energy at an average rate ϵ_d. As experimentally we observe that turbulence can persist in a (statistically)[5] stationary state, we have that $\epsilon_i = \epsilon_d = \epsilon$, which should also be the rate of energy transfer across the scales.

---

[4]For instance, if $u(x, t)$ is the velocity field at position x and time t, measuring, e.g., the spatial correlation means computing objects like $\langle u_i(x + r, t)u_j(x, t)\rangle$ and studying how they depend on r. Notice that in the above expression $\langle [\dots] \rangle$ denotes some form of average, which can be over different experiments or, if there are good reasons to assume that the properties do not depend on the specific point x, a spatial average.

[5]Which means that energy fluctuates around a well-defined mean value, i.e., it does not increase or decrease with time.

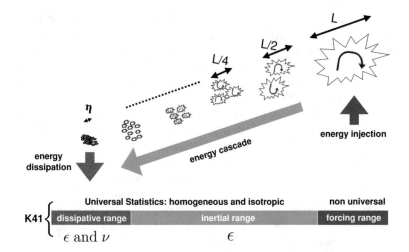

Fig. 27.2 Sketch of energy cascade. Large scale eddies, of size L in the order of the scale where energy is injected, become unstable and generate smaller scale eddies until they reach the scale η, where dissipation takes over. According to K41 theory, we can identify three ranges of scales: $r \sim L$ the (non-universal) forcing range, $\eta \ll r \ll L$ the (universal) inertial range where the statistics only depends on the energy dissipation rate ϵ and the dissipation range which is also expected to be universal and only depending on ϵ and the viscosity ν

The idea of developing a statistical theory of turbulence was further put forward by Andrei Nikolaevich Kolmogorov (see related entry) in 1941 in a series of works,[6] which still stand at the basis of much of our understanding of turbulence. At that time, Kolmogorov already laid the foundation of the axiomatic theory of probability, which explains the opening sentence of his first paper on turbulence *"In considering the turbulence it is natural to assume the components of the velocity [...] as random variables in the sense of the theory of probabilities"*. Then he made three hypothesis, whose basis have been discussed above: at scales small enough turbulence is homogeneous and isotropic and has universal properties; at very small scales such properties should depend only on the kinematic viscosity ν and the rate of energy dissipation ϵ; at small scales but not too small (i.e., where the energy cascade is at play) they should depend only on ϵ. Homogeneity and isotropy allow for disregarding the dependence on position and direction. Then, using dimensional arguments[7] we can realize that the small eddies responsible for dissipation should be at scale $\eta = (\nu^3/\epsilon)^{1/4}$ and with

---

[6]Note that in about the same years independently from Kolmogorov both Heisenberg and Onsager arrived, with different paths, to similar ideas. However, the formulation of Kolmogorov was more general and led to the modern understanding of turbulence.

[7]If $[L]$ and $[U]$ are used as dimensions for lengths and velocities, respectively, time can be expressed as $[L][U]^{-1}$. Then we notice that ϵ has the dimension of energy over time, i.e., $[U]^3[L]^{-1}$ while viscosity ν of square length over time, i.e., $[L][U]$. Then to make a scale using ν and ϵ one has $[L] = [\epsilon]^\alpha[\nu]^\beta = [U]^{3\alpha+\beta}[L]^{-\alpha+\beta}$, thus solving $3\alpha + \beta = 0$ and $\beta - \alpha = 1$ one obtains $\alpha = -1/4$ and $\beta = 3/4$, and analogously the other expressions.

characteristic velocity intensity $u_\eta = (\nu\epsilon)^{1/4}$. Finally, at intermediate scales, again dimensional arguments suggest that velocity increments (ignoring that the velocity field is three dimensional) $\delta_r u = u(r) - u(0)$ behave as $\delta_r u = (\epsilon r)^{1/3}$. This power law behavior should be universal within the range $\eta \ll r \ll L$, i.e., at scale enough larger than dissipative eddies (η) and smaller than the scale at which energy is injected (L), namely, in the so-called inertial range.

The theory of Kolmogorov (often dubbed K41) can also be used to understand in which sense turbulence is a very high-dimensional system. If we assume to discretize the space in a grid, clearly the minimal grid size should be at least of the order of the dissipative eddies η. Therefore, we will need at least $N = (L/\eta)^3$ grid points. Using the expression for η found by Kolmogorov and rearranging the terms, one can show that $N \sim Re^{9/4}$, considering that for a car Re can be as large as 10^7 one can understand what it does mean high dimensional. Summarizing turbulence is a strongly non-linear, high-dimensional non-equilibrium field theory, to the experts each of the adjectives sounds as intimidating.

Remarkably, K41 theory was in line with two experimental facts and with an exact result derived by Kolmogorov himself in 1941. Focusing on the experiments, the scaling proposed by Kolmogorov was compatible with early measurements of the power spectrum of turbulent signals (related to the second moment of velocity increments) and with the so-called "zeroth law" of turbulence. The latter amounts to the experimental observation that the energy dissipation becomes independent of the viscosity value for Reynolds numbers large enough, fact that can be derived from the $1/3$ scaling law. The zeroth law of turbulence (also dubbed dissipative anomaly) illustrates an important aspect of turbulence. For $\nu = 0$, the Navier-Stokes equations recover the Euler equations for an ideal non-dissipative fluid. However, the zeroth law states that in the limit $\nu \to 0$ energy is still dissipated, meaning that the limit $\nu \to 0$ is actually singular. While this seem as an exquisitely theoretical issue, it stays at the core of the impossibility to arbitrarily reduce the drag of a car, thus it is money! For ϵ to stay constant when $\nu \to 0$ it requires the gradient of the velocity field to diverge in that limit (as $\epsilon = \nu\langle|\nabla u|^2\rangle$), which implies deep mathematical problems at the basis of the fact that there is not yet a theorem of existence and uniqueness for the three-dimensional Navier-Stokes equations. This is a problem worth 1,000,000 $, according to the Clay Mathematics Institute.

Back to K41, the scaling behavior $\delta_r u = (\epsilon r)^{1/3}$ implies the following scaling for the moments $\langle(\delta_r u)^q\rangle \sim r^{q/3}$, which can be measured in experiments. For $q = 3$, this agrees with the exact result derived by Kolmogorov and for $q = 2$ it is in fairly good agreement with the experiments available in 1941. However, starting from the 1970s, refined experiments and also (later) numerical simulations were able to accurately measure higher order moments finding deviations from the predicted scaling. Essentially, while the behavior of the moments is still as a power law in the inertial range, i.e., $\langle(\delta_r u)^q\rangle \sim r^{\zeta_q}$, the exponents ζ_q display a non-linear dependence on q, and (for $q > 3$) are smaller than the K41 $q/3$ value. These deviations while small tend to increase at increasing the order q, for instance, $\zeta_6 \approx 1.78$ instead of 2. These deviations from K41 are dubbed *anomalous scaling* laws and express the intermittent nature of turbulent velocity as they imply that the velocity increments at small scales are much more intense than predicted by K41.

When discovered, many were believing that such deviations would disappear for very large Reynolds number. However, both experiments and numerical simulations convincingly showed that they persist at large Re and moreover they are, within the reachable accuracy, universal. This imposed to reconsider the K41 theory. The K41 scaling behavior means that in the inertial range velocity fluctuations are self-similar (the idea of self-similarity is also discussed in the entry Fractals). Pictorially, this means that an eddy of any size is uniformly filled by eddies of smaller sizes, until the dissipative scale is reached. Moreover, the small dissipative eddies should cover uniformly the whole fluid. However, simulations of turbulent flows show that this is not true (see Fig. 27.3, next page, showing the spatial organization of a quantity closely related to the energy dissipation). Indeed the non-linear dependence of the exponents ζ_q means a breaking of self-similarity.

At the beginning of the 80s, Parisi and Frisch proposed a generalization of K41 scaling idea to account for such experimentally observed deviations. In a nutshell, they conjectured that velocity increments at scale r possess a scaling behavior but with a exponent h, i.e., $\delta_r u \sim r^h$, over a set of fractal dimension $D(h)$ for an interval of values of h, hence the name of multifractal model. Moreover, they assumed that the dimensions $D(h)$ are universal for Re large enough. Then choosing appropriately the $D(h)$ one can fit the values of the exponents ζ_q and also make this generalized scaling compatible with the zeroth law of turbulence. Using the framework of the multifractal model, we can reinterpret K41 as the limit in which $D(1/3) = 3$, i.e., there is a single exponent $h = 1/3$ over a space filling set. In the multifractal model, self-similarity is broken: eddies of any given size will contain smaller eddies but in an inhomogeneous (not space filling) way. In practice, we have to think of a turbulent velocity field as a disorderly hierarchical organization of almost quiescent regions interrupted by strong fluctuations where dissipation takes place, in agreement with observations (as shown in Fig. 27.3).

A possible objection is that the multifractal model is a fitting model but not an explicative one. However, the multifractal model proved to have some explicative power. While one needs to fit the $D(h)$ from the measured values of the exponents ζ_q, one can use the multifractal model to make predictions on other quantities and then check whether the predictions are correct. As a matter of fact, over the years several predictions have been done and they were found to be consistent with (numerical or experimental) measurements. Thus, we can reasonably state that the multifractal model is, at present, the best statistical, phenomenological theory of turbulence. In order to remove the adjective "phenomenological" one would need to derive the $D(h)$ from the Navier-Stokes equations, which is a hard open problem.

In the last decade of 1900, anomalous scaling was understood in a closely related problem, namely, the transport of scalar field (like a dye or any other substance which is transported by the fluid without altering its velocity) by turbulent flows. The equations ruling the transport of a scalar field are much simpler than the Navier-Stokes equations as they are linear. Remarkably, when transported by a turbulent flow such scalar fields develop a cascade toward the small scales analogous to the

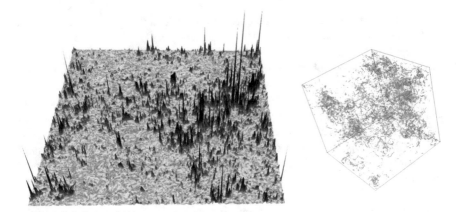

Fig. 27.3 Left: visualization of a 2D plane of enstrophy ($|\nabla \times u|^2$), closely related to the energy dissipation, extracted from a 3D simulation at a resolution of $N = 1024^3$ collocation points of a homogeneous and isotropic turbulent flow. Notice the coexistence of regions with low and high values denoting a non uniform distribution. Right: contour of the regions inside the 3D volume where the vorticity module is above 20% of its maximum ($|\nabla \times u| > 0.20 \max\{(|\nabla \times u|)$), again notice the non uniform filling of the volume. Courtesy of L. Biferale and M. Buzzicotti (see M. Buzzicotti, L. Biferale, and F. Toschi, Phys. Rev. Lett. **124**, 084504 (2020))

one described above and a K41-like theory can be derived. Also in this case, however, the scaling behavior is anomalous. Robert Kraichnan (1928–2008) in the mid-1990s proposed a very simple model in which the properties of the velocity field are under control arguing the presence of anomalous scaling. In the subsequent 10 years, the problem was exactly solved and the anomalous exponents were theoretically derived and shown to be universal and related to the existence of statistical conservation laws. While this was a great success, it is still unclear if and how such ideas can be exported to fluid turbulence, as they rely on the linear character of the equations while the Navier-Stokes ones are strongly non-linear.

After discussing the properties of turbulent flows, from an Eulerian point of view, it is useful to discuss it from the Lagrangian one, i.e., focusing on the motion of fluid elements (or small particles transported by the fluid). Indeed, one of the hallmarks of turbulence is its efficiency in transporting and mixing substances as highlighted by the following example. Your room mate lights up a cigarette say at 4 m from you, how long does it take for you to smell it? It is a matter of seconds, isn't it? Considering that smoke-parcels molecular diffusivity is about $D = 2 \cdot 10^{-5}$ m²/s and that for a diffusing particle (see the entry Brownian Motion), on average, the square of distance traversed is linear in the time duration, we would get about $8 \cdot 10^5$ s, i.e., slightly more than 9 days! Such big difference is because smoke spreading is mostly due to air turbulence and not diffusion. Indeed, even if not perceived, room air is turbulent due to air blowing in from windows or because excited by unavoidable thermal gradients. Turbulent spreading of smoke, odors, etc. is crucial for life: animals find their mates or their food mostly via odors; think of their life if odors only spread via molecular diffusion. Why is turbulence so efficient in spreading substances? We can identify main two mechanisms.

The first one is not specific to turbulence, as also velocity fields which are not turbulent can lead to a fast spread of substances over large scales. This is due to the fact that either for the chaotic nature of the velocity field and/or for a combination of transport by the flow (which can also be non-chaotic) and molecular diffusion, the velocity of the particles becomes uncorrelated over time and thus perform a sort of Brownian Motion but with diffusion coefficient which can be (again dimensionally) estimated as the product of the correlation time and the average of the square velocity, both of them can be large and typically their product is much larger (even by several order of magnitude) with respect to the molecular diffusivity.

The second one strongly relies on the properties of turbulent flows. Consider two particles which start close-by, say at distance R within the inertial range. Their distance will be controlled by their velocity difference which, according to K41 (assumed to be valid without corrections for the sake of simplicity), scales as their distance to the power $1/3$. Thus, the separation between the two particles will evolve, roughly, as $dR/dt = R^{1/3}$ which implies $R^2 \sim t^3$. Recalling that for diffusing particles $R^2 \sim t$, we can understand that the spread of a cloud of initially close particles will be explosive in turbulence as beautifully illustrated in panel (a) of Fig. 27.4. In modern research, such important deviations from diffusive behaviors are termed anomalous diffusion. Remarkably, in 1926 (thus before K41 theory), Richardson, examining data from atmospheric balloons, deduced that their distance tends to grow on average as t^3. Moreover, he proposed a theory for this in terms of a diffusive equation with a scale-dependent diffusion coefficient of the form $D(R) \propto R^{4/3}$. Notably, using K41 framework assuming the diffusion coefficient to depend only on the scale R and the energy dissipation ϵ, one would dimensionally derive $D(R) = \epsilon R^{4/3}$, as proposed by Richardson.

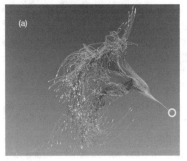
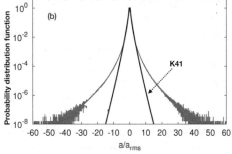

Fig. 27.4 Lagrangian turbulence: **a** A bunch of small particles initially at small distance between each other, within the yellow circle, separate explosively in a turbulent flow; **b** Probability distribution function of Lagrangian acceleration, the black solid curve shows the prediction derived using K41 theory, the multifractal prediction (not shown) basically superimposes onto the data (red curve). Courtesy of L. Biferale (see also L. Biferale et al. Phys. Rev. Lett. **93**, 064502 (2004) and R. Scatamacchia et al. Phys. Rev. Lett. **109**, 144501 (2012))

In the last two decades, there have been experimental advancements in the ability to study turbulent flows also from a Lagrangian point of view, which allowed to measure the instantaneous velocity and acceleration of very small particles in high Reynolds number flow. In parallel, also large-scale simulations seeded with millions of particles have been performed. These studies revealed very interesting properties. For instance, what we discussed above about the statistics of velocity spatial increments can be repeated for temporal increments of the velocity. It is possible to make predictions on moments of such temporal increments by using the multifractal model, after translating it from the Eulerian to the Lagrangian frame. Remarkably, experiments and simulations have shown to be largely compatible with such predictions. Similarly, this has been done for the particle acceleration (which is nothing but the left-hand side of the NS equation) probability distribution function (see panel (b) of Fig. 27.4). The figure also shows a remarkable qualitative property of the acceleration: a quite large probability to find events in which the acceleration is 60–80 times the typical (root mean square) value. A closer inspection revealed that such events of extremely large acceleration are typically associated to vortex tubes, i.e., vortices with a core on the scale of the dissipative scale but which extends on much longer scales and persists for some time. One can think of them as micro-tornadoes. To give an idea of the implications, in a day with a 20 km/h wind, there are mini-tornadoes in which the acceleration can be up to a hundred time the gravitational one (i.e., 100 g). A man can sustain accelerations larger than 10 g only for a few seconds. However, we can relax, as such extreme events happen on so small scales that we cannot even perceive them. Fortunately for us, however, such scales are perceived by mosquitos, which indeed tend to disappear when there is a strong wind, or just a ventilator in our room.

We wish to conclude this brief tour through turbulence research with a few general considerations. Usually when people come to know that someone is a physicist the first asked question is: are you studying particle physics or cosmology, what new particle or law has been discovered? This is a pretty natural consequence of the fact that, especially at the level of science popularization and unspecialized media, physics breaking news typically concerns the discovery of new particles, cosmological phenomenon, etc. We know the law regulating fluid motions since about 200 years, and there are very good reasons to believe that they are correct and not necessitating a revision. However, as it should be clear from the above survey, we still not have a detailed understanding of turbulence. Thus, turbulence is a perfect example of how the knowledge of the "laws" regulating a given phenomenon can just be the beginning of the scientific knowledge of that phenomenon. This is perfectly expressed by Feynman in the first volume of his lectures (1963) and it is still (with some caveats) valid nowadays: *"Finally, there is a physical problem that is common to many fields, that is very old, and that has not been solved. It is not the problem of finding new fundamental particles, but [...] It is the analysis of circulating or turbulent fluids. [...] The simplest form of the problem is to take a pipe that is very long and push water through it at high speed. We ask: to push a given amount of water through that pipe, how much pressure is needed? No one can analyze it from first principles".*

The unavailability of an exhaustive and complete understanding of turbulence, joined with the fact that many giants of science have approached the problem without a sharp success, is typically popularized saying that turbulence is a mystery[8] and a cemetery of theories. While we cannot deny the succession of incomplete or wrong theories proposed for turbulence, we wish to point out that, in this still unfinished chase for a theory of turbulence, many fruitful ideas and tools have been developed. For instance, the ideas of scale invariance and universality have been very fruitful in the statistical mechanics of critical phenomena and in quantum field theory. Richardson dispersion has stimulated the whole field of anomalous diffusion. The multifractal formalism has found applications from the analysis of financial data to medical imaging, from the traffic over Internet to neuroscience. Finally, interestingly from an epistemological viewpoint, the use of computational techniques in turbulence has an almost unique status: solving the Navier-Stokes equations with a computer representing an in silico experiment which can be compared with laboratory experiments. This was already in the mind of one of fathers of electronic computers: as early as 1949 John von Neumann (see the corresponding entry) wrote *"[...] there might be some hope to 'break the deadlock' [of turbulence] by extensive, but well-planned, computational efforts. It must be admitted that the problems in question are too vast to be solved by a direct computational attack [...] There are, however, strong indications that [...] relevant information must be obtained by direct calculations"*, and he was right.

---

[8] *"When I meet God, I'm going to ask him two questions: why relativity? And why turbulence? I really believe he'll have an answer for the first"*. This probably apocryphal quote was attributed to Heisenberg and by others to Lamb.

Chapter 28
Universality

In 1936, Einstein wrote *"the eternal mystery of the world is its comprehensibility. [...] The fact that it is comprehensible is a miracle"*. While our understanding of the basic laws of Nature is rather incomplete, we cannot deny that many (even complex) phenomena have been rationalized in spite of our fragmented knowledge. In this entry, we shall discuss a few examples in which, luckily, it is possible to derive, at least partially, interesting results even without a complete control on the investigated problem.

As a first instance of this possibility, we like to mention dimensional analysis whose use, as already established in the nineteenth century, by J. C. Maxwell, allows for obtaining useful results from simple arguments. A pedagogical example, usually taught in the first year of physics and engineering courses, is as follows. Consider a pendulum of length ℓ, with a mass m at its end, in a gravitational field with acceleration g. It is a common experience (think of the pendulum clock) that its motion is periodic. Clearly, the period T_p should depend on m, g, and ℓ, can we say how without solving Newton's equation? Dimensional analysis tells us that we should obtain a [time] (T_p) using quantities whose physical dimensions are [mass] (m), [length]/[time]^2 (the acceleration g), and [length] (ℓ). It is an easy exercise to find that the unique combination is $m^0 g^{-1/2} \ell^{1/2}$ so that $T_p = C\sqrt{\ell/g}$. Here C is a constant without physical dimension, whose exact value, yes, requires solving the Newton equation. The key facts we can obtain via dimensional analysis are that the value of the mass m is not important, while ℓ and g appear only in the combination $\sqrt{\ell/g}$. Solving Newton's equation we discover that the constant C depends on the value θ_0 of the angle at the initial time and that for small values of θ_0, $C \approx 2\pi$. This result is deceptively simple; however, dimensional analysis can be used to obtain much deeper knowledge.[1]

---

[1]For instance, consider the problem of the black-body radiation. Around 1860 Kirchhoff proved that the (electromagnetic) energy density of a black body should be a universal function of the temperature. In 1884, Boltzmann proved that the energy density is indeed proportional to the

A keyword for our success in the understanding of (at least some) complex phenomena of our world is *universality*: such a term had been introduced in the 1960s by theoretical physicists in the context of the critical phenomena, among the many we cite Widom, Domb, and Kadanoff. The basic idea, however, traces back its origin to probability theory.

As non-trivial examples of universality we recall two very important results of probability: the law of large numbers (LLN) and the central limit theorem (CLT). In a few words, LLN and CLT deal with the properties of the empirical mean $y_N = (x_1 + x_2 + \cdots + x_N)/N$ of N random variables, $\{x_1, x_2, \ldots, x_N\}$. In the limit case of N very large, LLN and CLT give an answer to two natural questions: the mean behavior of y_N and the statistical features of small fluctuations of y_N around its mean value. In the case of sequences $\{x_1, x_2, \ldots, x_N\}$ of independent and identically distributed random variables with mean value m and finite variance σ^2, the LLN answers the first point. The empirical average gets closer and closer to the expected value m when N is large. The second point is addressed by the CLT. In the simple case of independent and identically distributed variables with expected value m and finite variance σ^2, the CLT tells us that in the limit $N \to \infty$, the quantity $z_N = (\sqrt{N}/\sigma)(y_N - m)$ is distributed according to the bell-shaped curve of Gauss $e^{-z^2/2}/\sqrt{2\pi}$. Under suitable (and rather general) hypothesis, the above results can be extended to dependent (weakly correlated) variables.

Let us underline the important conceptual aspects: in the above results only the mean value m and the variance σ^2 appear. This is a manifestation of some form of universality: it is not necessary to have a very precise knowledge of how the random variables $\{x_n\}$ are distributed, only m and σ^2 matter, the details on the distribution of the random variables are irrelevant and do not appear in the result. As discussed below, it is this independence of the details which makes universality a powerful concept allowing to derive important results without exact knowledge of all the particulars.

Thermodynamics represents a field of physics where strong results exist which do not depend on several aspects of the system. The concept of Carnot efficiency is a famous example (see also the entry Irreversibility). Sadi Carnot (1796–1832) in his book on thermodynamics *Réflexions sur la puissance motrice du feu et sur les*

---

fourth power of the temperature—the well-known Stefan-Boltzmann law. Only in 1900 Max Planck heuristically derived the expression for the energy density per frequency, giving birth to quantum mechanics. Now let us see the power of dimensional analysis: energy density, i.e., energy per unit of volume, has the physical dimension of [mass]/([length][time]^2). From Kirchhoff's universality, using temperature and universal constants we should be able to derive the correct law but for a non-dimensional constant. At the beginning of 1900, the only available physical quantities with the above characteristics were $k_B T$ (where k_B is the Boltzmann constant) which has the dimension of [energy]=[mass][length]^2/[time]^2 and the speed of light c with dimension [length]/[time], which enter Maxwell's equations of electromagnetism. The reader can try but only with these two quantities it is impossible to obtain the physical dimension of an energy density. Hence, there must be another universal constant from a "new" theory. This is Planck's constant h which has the dimension of [mass][length]^2/[time]. Using h, $k_B T$, and c one can indeed derive that the energy density should be equal to $C (k_B T)^4/(c^3 h^3)$, and a refined theory gives the constant $C = 8/(15\pi)$.

machines propres à développer cette puissance[2] gives a very general description of the physical principles involved in real steam machines, proposing a sort of universal heat engine which incorporates the minimal ingredients to transform heat into work. Those ingredients are: a heat source (fire); a heat sink (air); their temperatures T_h and T_c, respectively (obviously with $T_c < T_h$); the heat taken from the source; the heat (inevitably) dissipated into the sink; and work extracted, which is the difference between the two heats. In terms of this machine, Carnot was able to state a primitive form of second principle of thermodynamics: the efficiency of an engine cannot be greater than $1 - T_c/T_h$. All details of a real engine (volume, chemical substances involved, loads applied, etc.) do not enter this very important formula.

A further example is found in the dynamics of fluids. For all practical purposes, the Navier-Stokes equations which rule the time evolution of the fields (velocity, pressure, etc.) can be (and were originally) derived applying Newton's laws and thermodynamics to a continuous medium. On the other hand, at least in some cases (dilute gases), such equations can be rigorously obtained starting from the microscopic description, that is, considering a fluid as a large collection of interacting molecules, whose evolution is governed by the classical laws of mechanics. Such a procedure is surely much more fundamental than the continuous medium approach. The history of science suggests that this approach from micro to macro level is not really necessary: how could Euler and Bernoulli, in the eighteenth century, have understood hydrodynamics so well that their results still hold today? They had no knowledge at all of the microscopic dynamics, since at their time there was no way of knowing how molecules interact. The reason is that for the derivation of the Navier-Stokes equations not all the details of the microscopic deterministic dynamics are important, but only global properties (mainly the conservation laws of mass and momentum) are relevant.

Therefore, in order to understand a complex phenomenon such as turbulence, it is not necessary to follow the motion of the atoms in a liquid, it is enough to solve the Navier-Stokes equations. Such a mathematical problem, because of the presence of non-linear terms in the equations, is a nightmare. In spite of these formidable difficulties, today we have a rather good understanding of the statistical properties of the turbulent flows at small scales. The reasons of this success had been the intuition of L. F. Richardson (see related entry) on the cascade mechanisms and the idea of A. N. Kolmogorov on small-scale universality (see the entry Turbulence for details). Richardson suggested that, in a turbulent flow, the largest eddies are created by instability of the mean streamline. Then these big eddies, with a similar mechanism, generate smaller eddies (say half their size), and so on, generating a cascade down the smallest eddies at a viscous scale (now called the Kolmogorov length), where energy is eventually dissipated. In 1941, Kolmogorov, in two celebrated short papers, was able to understand that in the so-called inertial range, i.e., at scales much smaller of where the fluid is stirred and much larger of where energy is dissipated (see

---

[2] Available without any access fee at www.numdam.org.

the entry Turbulence), the only relevant parameter, besides the obvious scale of the eddy, is the rate of energy consumption. Then he derived new crucial results mainly using dimensional analysis. As far as we know this was the first non-trivial case of universality in physics, the statistical properties at small scales are independent of the details of the stirring mechanism (i.e., the external pump of energy) as well as of the geometry of the container and the fluid viscosity.

Let us now briefly discuss critical phenomena, a topic which may appear rather technical, but which is conceptually very important. Indeed, in the 1960s–1970s, the theoretical and experimental works on this topic conspicuously demonstrated that a variety of macroscopic phenomena do not depend on microscopic details and thus put forward the concept of universality.

Consider the phenomena known as second-order phase transitions where one has a continuous change of certain material properties, as example we can cite the case of the liquid-gas and the standard fluid-superfluid transition. These transitions are characterized by the presence of critical points in the phase diagram for equilibrium states (for instance, the liquid and gas phases). It is customary to describe liquid-gas systems in terms of an equation of state, namely, a thermodynamic relationship involving the temperature T, pressure P, and density ρ usually in the form $F(P, T, \rho) = 0$.[3] Usually we can represent the equation of state in terms of a graph in the (P, T) (or (P, ρ) see right plot) plane, as in Fig. 28.1. Typically, for given values of (P, T) (or P, ρ) the system is either liquid or gaseous (or solid); however, there are lines in the plane where two phases can coexist. As shown in the figure, the liquid-gas co-existence line does not extend forever, it stops at a critical point, of critical thermodynamic coordinates P_c, ρ_c, T_c around which the liquid phase turns continuously into gas and the distinction between the two phases fades away.

The liquid density ρ_l is larger than the gas density ρ_g and the difference $M = \rho_l - \rho_g$ (in technical jargon called the order parameter) vanishes when T tends to T_c in the following way:

$$M \sim t^\beta \quad \text{with,} \quad t = \frac{T - T_c}{T_c},$$

where t and β are usually called the reduced temperature and the "critical" exponent, respectively. Actually, a few other critical exponents can be introduced to describe the behavior at small values of t of other interesting physical quantities such as the specific heat.

The most striking feature of critical phenomena is that the critical exponents, such as β, are largely independent of the details of the microscopic dynamics of the elementary constituents of the system of interest, i.e., they are "universal". Experiments show that a wide range of even very different systems, with very different critical temperatures, share the same critical exponents. For instance, strikingly, the critical

---

[3]For instance, for the ideal gas the equation of state is the well-known $P - \rho(R/M)T = 0$ where R is the gas constant and M the molar mass (in Kg per mole).

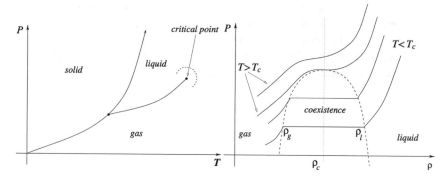

Fig. 28.1 Graphical representation of the solid-liquid-gas phase diagram (left) and of the equation of state (right). Notice that for $T \gg T_c$ the equation of state should tend to that of an ideal gas (see footnote 3)

exponent of the liquid-gas transition is the same of the paramagnetic-ferromagnetic transition.[4]

Naturally, not all known critical phenomena are characterized by the same exponents, but they can be grouped into just a few "universality classes", within each of which the exponents do not vary. Remarkably these classes depend only on very few parameters, usually the dimensionality and the symmetries of the systems of interest, and not on the details of the microscopic rules controlling the interaction between the system components.[5]

A posteriori we can say that the existence of few classes of universality, depending on a limited number of parameters, constitutes a great blessing for science. In the absence of such a lucky property, understanding the ferromagnetic transition would require a very detailed control of the microscopic interactions of each specific material (the shape of the Hamiltonian in the physical jargon), something far beyond our real possibilities. On the contrary, it has been possible to derive quantitative results in agreement with experiments by studying very simplified models such as the celebrated Ising model. In such a model, the ferromagnet is represented as a lattice of discrete variables taking values 1 (up state) and -1 (down state) and interacting in such a way that the energy is decreased when close-by lattice sites are both up or down, and subjected to thermal fluctuations which tend to disorder the up-down state. Such an example is particularly interesting as it points out that, at least in some cases, it is not necessary to build fully detailed descriptions of a system (we let the reader to imagine the complexity of the interactions between the atoms composing a

---

[4]The critical temperature for ferromagnetism is called Curie temperature, and it turns out that for many materials T_c is much higher than the ambient temperature so that they behave as permanent magnets. When the temperature is raised above T_c they become paramagnetic and loose the permanent magnetization.

[5]Which are typically described in terms of an energy function called Hamiltonian, e.g., for a gas-liquid the Hamiltonian is the sum of the kinetic energy of the constituents and the interaction potential among them.

ferromagnet) to obtain even delicate quantitative results (i.e., the critical exponents). This short discussion should have convinced the reader about the astounding (and a priori unexpected) fact that models—at least in physics—do not need to adhere precisely to reality to be useful (see the entry Laws, levels of descriptions, and models for a further discussion on modeling reality).

Chapter 29
Volterra

Vito Volterra (1860–1940) has been a mathematician who played a very important role in the progress of scientific institutions as well as in the organization of science. He gave seminal contributions to many fields of mathematics, in particular, in integral equations, functional analysis, and modeling of biological systems. Volterra was very interested in technology and applications of mathematics; we can mention his contributions to the theory of dislocations in crystals, hereditary phenomena (i.e., systems with memory), and overall models of biological systems.

Volterra was born in Ancona, at that time part of the Papal States, into a poor Jewish family, student at the Scuola Normale in Pisa, he graduated in physics in 1882. His career was very brilliant and fast: full professor of mechanics in Pisa at age 23, then in Turin (1892), at the end professor of mathematical physics in Rome (1900). In 1905 he was named by the king Victor Emmanuel III as a senator of the Kingdom of Italy.

He had a significant public profile with a very active role in the political and cultural Italian life, in particular, in the organization of science. He proposed (1906) to found an Italian Association for the Advancement of Science (*Società Italiana per il Progresso delle Scienze*).

During World War I, Volterra (who was volunteer for the Italian army) had an important role as Head of the Office of Inventions, created to coordinate the military, industries, and universities. The experience of the war and its outcomes inspired to Volterra the idea of promoting the foundation of the *Consiglio Nazionale delle Ricerche—CNR* (National Research Council) to coordinate and direct applicative researches and facilitate their use for the technological and economic progress. He was the first president of CNR (1923), but, because of his opposition to fascism, he was not reconfirmed.

© Springer Nature Switzerland AG 2021
M. Cencini et al., *A Random Walk in Physics*,
https://doi.org/10.1007/978-3-030-72531-0_29

In the first decades of the twentieth century, Volterra had a rather influential role in the organization of Italian scientific community, even beyond his specific field of activity, for instance, in 1912 he became the first president of the Comitato Talassografico Italiano for the marine research and was involved in the establishment of the national meteorological network. In 1902, he affiliated to the Institute of Physics of Rome, which, under the direction of Pietro Blaserna, wanted to promote the development of modern physics in Italy. Volterra supported the researches in nuclear and theoretical physics, with his help the rising star Enrico Fermi obtained a Rockefeller scholarship and then the first Italian chair for theoretical physics (1926).

Volterra, as many eminent Italian scientists of his time, followed the risorgimental liberal tradition, and was a strong opponent of the fascist regime of Mussolini. In 1923, he openly opposed to the educational reform of G. Gentile, in 1925 he signed the *Manifesto on the intellettuali antifascisti* (Manifesto of anti-fascist intellectuals) proposed by B. Croce. His political philosophy is well summarized by a post card he sent in 1930s *"Empires die, but Euclid's theorems keep their youth forever"* (see Fig. 29.1), an elegant criticism of the imperialistic Mussolini's regime.

Although he was both senator and the most eminent Italian scientist, he was under the constant control of OVRA (the fascist secret police). In 1931, he was one of the few professors of the Italian universities (only 12 out of 1250) who refused to take a oath of loyalty to the fascist regime. As a consequence he was compelled to resign his university position and, after the infamous antisemitic legislation (1938), also his membership of scientific academies. He spent his last years largely abroad and in Ariccia (a small town close to Rome). *"This morning at 4:30 in his home in Via in Lucina 17, the Senator Vito Volterra son of Abram, of Jewish race, died"*: this was the coarse report sent (on October 11, 1940) by the police to the Ministry of the Interior to inform them that the dossier being kept on Vito Volterra could finally be closed.

Volterra had been a towering personality of Italian science (four times plenary speaker in the International Congress of Mathematicians, co-founder of the Italian Physical Society, first President of CNR, President of Accademia dei Lincei, etc). In spite of this the Pontificia Accademia delle Scienze was the only Italian institution that organized a solemn ceremony to remember the great scientist just after his death.

In 1901, Volterra gave the inaugural lecture of the academic year at the University of Rome, discussing some attempts at applying mathematics to biological and sociological sciences (*Sui tentativi di applicazione delle matematiche alle scienze biologiche e sociali*). This lecture circulated widely, it was translated in French and both the Italian original and the translation were repeatedly reprinted, and had an influential role in the history of applied mathematics. We can say that with this conference Volterra had been one of the first important scientists who wondered in a serious way about the role of mathematical methods in contexts different from physics or chemistry.

Vito Volterra was also interested on some data from commercial fishing in the period 1903–1923 that were shown to him by his future son in law U. D'Ancona. The data concerned the cartilaginous fish in the catch of three Adriatic ports, Trieste,

Fig. 29.1 Portrait of Vito Volterra with the famous sentence *"Empires die, but Euclid's theorems keep their youth forever"*. ©Volterra family archive, Villino Volterra, Ariccia, Italy

Venice, and Fiume (now Rijeka). D'Ancona observed a clear growth in the proportion of these species during the World War I, a period of very little fishing. Such a rather counter-intuitive behavior was at the origin of the interest of Volterra in the building of mathematical models for biological phenomena. In the building of his model Volterra reasoned by analogy with the kinetic theory: *"the big fish collide with the small fish and with a certain probability the former eats the latter"* and thus came to find a set of two differential equations for the time evolution of the predator and the prey populations.

The Lotka-Volterra model was (independently) proposed by A. J. Lotka in the theory of autocatalytic chemical reactions in 1910. Before them the use of mathematics in biology did not amount to a great deal. The first attempt dates back to Fibonacci with his celebrated sequence which had been originated from a puzzle about the growing of a rabbit population model. After this pioneering contribution, the modern foundation of population dynamics is due to T. R. Malthus, then we have K. Pearson's biometric research, P. F. Verhulst's studies on the logistic equation and the mathematical model introduced by R. Ross to describe malaria epidemics.

With Volterra the role of mathematics in biology changed: no longer an ancillary tool of limited relevance in assisting the analysis of biological phenomena, but a source of new concepts and methods. In the words of Volterra in his inaugural lecture in 1901 *"the translation of natural events into the language of arithmetic or geometry means opening up to mathematics rather than an exercise of the analytical tools"*.

The celebrated model of Lotka and Volterra opened the *Golden Age of the Theoretical Biology*: between 1920 and 1940 the fields of population genetics and mathematical theory of epidemics had an impressive growth. Once the Lotka-Volterra model has been constructed, we can reason about it in a purely mathematical way, asking ourselves if it is possible to extend the expressivity and, therefore, the predictive ability of the model itself. Remarkably, such a model, in spite of its simplicity, is able to describe several non-trivial behaviors in ecological systems in agreement with observations, e.g., the lynx and snowshoe hare data of the Hudson's Bay Company. Moreover, the Lotka-Volterra model found applications in economics theory.

It is worth mentioning that in ecology there is nothing similar to Newton's mechanics or Maxwell's equations, therefore the building of evolution model for ecological systems cannot be sought by relying on first principles. There are just few general assumptions which can be invoked to write down ecological models, e.g., to keep some analogy between the Lotka-Volterra equations and the consistency with the relevant ecological facts. On the other hand, in spite of the apparent poorness of the model, in particular, if compared with those in physics or astronomy, now it has well established its seminal relevance for its ability to predict qualitative features.

The elegance of Volterra's reasoning stands in contrast with the plausibility of his equation which can be criticized from several point of view. We mention, among the many, two important contributions which can be seen as a continuation of the Volterra ideas. In 1936, Kolmogorov, doubtlessly in honor of Volterra, published a short note in Italian, considering a simple and intriguing dissipative generalization of the Lotka-Volterra model of an ecosystem with just two species: prey and predator. In 1976, S. Smale (Fields Medal in 1966) gave an important contribution to the mathematical properties of systems of differential equations describing N competing species.

Chapter 30
von Neumann

John von Neumann (1903–1957) is one of the very few modern scientists who, although not universally famous, have profoundly impacted the development of various theoretical and applied disciplines. His intuitions and ideas basically shaped the computer technology as we know it today. Concerning pure mathematics, the breadth and the depth of von Neumann's works is definitely outstanding, surely he is the greatest scientist of the last century who has been able to produce excellent theoretical and applied mathematical results, which are considered fundamental milestones and have led to the creation of new disciplines, e.g., operator algebras or game theory, and have given a significant acceleration to technological development, e.g., for the structure of modern computers.

Born into a wealthy Hungarian family, Neumann János Lájos, this is his name at birth, was a child prodigy. At age of 6, he was able to converse in ancient Greek and to make complex mathematical calculations in his head. As it was common in the cultured and wealthy Hungarian families of that time, the young János received private education, in his case enhanced in the study of languages and mathematics. At age of 10, János enrolled in the most prestigious Gymnasium of Budapest, one of the most fertile environment in the entire Europe at that time. János' math teacher, Lazlo Ratz, was one of the foremost responsible of the reform of teaching mathematics, which was one of the elements that allowed the Hungarian miracle, namely, the simultaneous appearing, within a few years, of a large number of prominent mathematicians and physicists, among which we mention, besides János, Leó Szilárd, Eugene Wigner (Nobel prize in 1963), Edward Teller[1] and Paul Erdös. An extremely stimulating environment combined with his extraordinary capabilities allowed the young János to become one of the most important scientists of the twentieth century. Shortly after the beginning of the gymnasium, Lazlo Ratz recognized von Neumann's genius and suggested his father to have him take private mathematical lessons in order to make

---

[1] All these four scientists played a key role in the the Manhattan Project which led to the development of the atomic bomb.

© Springer Nature Switzerland AG 2021
M. Cencini et al., *A Random Walk in Physics*,
https://doi.org/10.1007/978-3-030-72531-0_30

the most of his natural talent and expand his mathematical knowledge. At the end of the gymnasium, von Neumann participated and won the Eötvös math competition and that was probably the moment in which he became fully aware of his exceptional mind.

Looking for a common thread in von Neumann's vast and varied scientific production, his most important source of inspiration surely was the interest in axiomatic formalization. Starting from the axiomatization of set theory, passing through that of quantum mechanics, not to mention his approach to game theory or the seminal contribution for the development of modern computers, throughout his scientific career his interests have often focused on formalization problems. Already his second scientific paper—*Zur Einführung der transfiniten Ordnungszahlen* (Toward the Introduction of Transfinite Ordinal Numbers)—prepared in 1921, when he was still in high school, was about formal problems. In this work, using words that already denoted a confident and mature style, the 18-year-old János stated: *"The aim of this work is to consider concretely and precisely the idea of Cantor's ordinal numbers [...] it is necessary to turn the somewhat vague formulations of Cantor himself into more precise definitions"*.

After graduation he enrolled in the math faculty at Budapest University and, pushed by his father, who insisted on a more practical and profitable preparation, he started to study chemical engineering at the Federal Polytechnich of Zurich (ETHZ). In this period, he spent most of his time in Zurich and returned to Budapest only at the end of the semester to take the exams of his course of study in mathematics (without attending the lessons). In Zurich, he also spent some time at the Department of Mathematics interacting with two great mathematicians, Hermann Weyl and George Polya. When Weyl had to leave Zurich for a period, von Neumann replaced him for teaching his course.

In 1926, in a few months, first he graduated in chemical engineering at ETHZ and then he discussed his final dissertation in Budapest as a candidate for an advanced doctoral degree in mathematics, which was based on his third work on the axiomatization of Cantor's set theory. The incipit of his paper was in perfect von Neumann's style *"The aim of the present work is to give a logically unobjectionable axiomatic treatment of set theory"*, denoting how von Neumann was becoming increasingly confident. To appreciate the difficulties of the argument, we recall that the axiomatization of Georg Cantor's set theory was one of the most debated subjects in the mathematics of the beginning of the last century and it had already put great mathematicians in difficulty. In his fundamental work about set theory, *Die Axiomatisierung der Mengenlehre* (The Axiomatization of Set Theory, 1928), an extension of his dissertation thesis, with the redefinition of two concepts, sets and classes, he canceled from set theory those pathological monsters which were yet leading to contradictions, such as the sets of all sets present in the Zermelo-Fraenkel axiomatization of set theory. The work of von Neumann on set theory was later used and completed by Kurt Gödel, and nowadays the axiomatization of set theory by von Neumann-Bernays-Gödel (NBG) is still used.

His interests in the foundations of mathematics had a perfect interlocutor, the great David Hilbert. Unlike other prominent mathematicians of the time, such as

Luitzen Brower, founder of the intuitionism in mathematics, for whom some parts of set theory were too unrigorous to be safely used at all, or Hermann Weyl who about Cantor set theory was thinking *"such dealings with infinity were trafficking with the devil"*, Hilbert thought that with a solid axiomatic basis for any topic (including set theory) the mathematics would progress mightily. In his very famous list of 23 unsolved mathematical problems posed in 1900, in the first two positions Hilbert posed two fundamental problems of axiomatization, "The continuum hypothesis" and "Prove that the axioms of arithmetic are consistent", and in the sixth position there is another important axiomatization problem "Mathematical treatment of the axioms of physics".

After graduation and with a grant from the Rockefeller Foundation (his first link with the United States), von Neumann went to the university of Göttingen, where he was awaited with open arms by Hilbert who, having already met him in Göttingen in 1925, was considering him one of the most promising among the young mathematicians of that period. In Göttingen, von Neumann found an incredibly fertile and stimulating environment, with all the leading scientists of the time trying to lay a solid foundation for quantum mechanics, the most important problem in those years. A whole series of physical phenomena seemed to involve transfers of energy in discrete packets (quanta) and not with continuity as expected from classical physics. From black-body radiation[2] to the photoelectric effect,[3] just to cite two major examples, there were clear evidence that a new physics was needed.

In those years, two apparently incompatible ways to approach quantum mechanics were proposed. The first, introduced by Werner Heisenberg (1925), was based on matrices representing the physical state of a given system. The second, put forward by Erwin Schrödinger (1926), was based on wave functions, which, once squared, provide the probabilities of finding a physical system in a given state. The axiomatization of quantum mechanics was of fundamental importance for David Hilbert as well as for the whole scientific environment of Göttingen, and seemed to be a problem perfectly tailored for von Neumann. Indeed, even if there had already been some attempts to unify the two interpretations, the formalism was not satisfactory. In a first paper written together with Hilbert and Nordheim, von Neumann began to build a solid mathematical apparatus at the basis of quantum mechanics, and, in subsequent works, in which he worked alone, he formulated a complete axiomatic description of quantum systems starting from the elements of a Hilbert space. The works of those years culminated in his first book *Mathematische Grundlagen der Quantenmechanik* (1932) (Mathematical Foundations of Quantum Mechanics) which is considered, even today, one of the most solid and mathematically rigorous book on

---

[2]That is the emission of radiation from a perfectly absorbing body at a given temperature. The first solid theory able to correctly describe the whole spectrum of the black-body radiation was proposed by Max Planck who imposed a discretization in the exchange of energy, i.e., the energy can be exchanged only in discrete packets multiple of the base unit $h\nu$, where h is Planck's constant and ν is the frequency of the electromagnetic wave.

[3]The photoelectric effect is about the emission of electrons when radiation hits materials. This effect was successfully described by Albert Einstein considering radiation as composed of discrete quanta, now called photons, based on Max Planck's theory of black-body radiation, see entry Einstein.

quantum mechanics. Remarkably, his mathematical formulation survived to the two major important developments of quantum mechanics after the book was written, namely, relativistic quantum mechanics and quantum field theory.

In those years, von Neumann also wrote his first paper on game theory, a topic more applied (related to economics) than to those pure theoretical subjects that had characterized his previous production. Game theory was attracting the interest of other important mathematicians of the period, as Ernst Zermelo, who presented a theorem about zero-sum games (a game in which the winning of a player is equal to the loss of the other) which does not involve chance, or Emil Borel who, again in the context of zero-sum games, introduced the formalism of the payoff matrix, that is still used today. In this context, von Neumann formally defined what games were referred to, what the rules and what the possible results were, and then he proved the minimax theorem, i.e., the existence of a strategy that allows to minimize the maximum losses. For about 20 years such a result has almost not been used until the publication of the book *Theory of games and economic behavior* (1944) by von Neumann and Morgestern, where they provided an axiomatic theory of expected utility and gave a rigorous mathematical basis to many economic problems, allowing also to treat decision-making process under uncertainty in a rigorous context. In that period, after being used by the British navy during World War II, von Neumann's games theory became a fancy topic of research. In Princeton, during the years in which von Neumann and Morgestern were present, another genius mathematician, John Nash, in his doctoral thesis extended game theory by introducing non-cooperative games and demonstrated that at least one equilibrium solutions, i.e., those solutions in which no player would benefit from changing their strategy, must exist. From that moment, any social, political, or economic theory involving the interaction between individuals each of which pursues its own objective, whether selfish or altruistic, has its theoretical reference model in game theory.

Before leaving Göttingen, von Neumann witnessed an earthquake that undermined the foundations of the axiomatic formalization of mathematics. In the conference "Epistemology of the Exact Sciences" held in Könisberg in September 1930, a very young and unknown mathematician—Kurt Gödel—presented a lecture about the incompleteness of the axiomatic method. He showed that, an axiomatic system including at least basic arithmetic is incomplete, meaning that there exist true propositions about natural numbers that cannot be proven within the axiomatic system. von Neumann, who was participating in the conference to present the thoughts of the formalist school as developed by Hilbert, immediately realized the revolutionary implications of Gödel result, and spent the following months elaborating these new ideas. While Gödel results surely frustrated his confidence on the Hilbertian formalist program, he never lost confidence on the importance of the axiomatic method, in his words *"The main hope of a justification of classical mathematics—in the sense of Hilbert or of Brouwer and Weyl—being gone, most mathematicians decided to use that system anyway"*. Possibly an extreme synthesis of von Neumann's mature attitude toward mathematics is epitomized by his famous sentence *"In mathematics you don't understand things. You just get used to them"*—it is not important to find meaning in mathematical objects, what is important is to rigorously define objects and rules in order to obtain solid theoretical or practical results.

In those years, von Neumann was started to be recognized as one of the most important scientists of the period. He received the teaching qualification in 1927 and became a Privatdozent at the University of Berlin in 1928, becoming the youngest teacher ever in the history of the Berlin University. As reported in the book of Norman Macrae on von Neumann *"By the end of 1927 Johnny had published twelve major papers on mathematics. By the end of 1928 these had risen to twenty-two, by the end of 1929 to thirty-two. They had established something of a cult among at least Europe's younger mathematicians, who through 1928–1929 were avidly awaiting his extraordinary output of nearly one major paper a month"*. From 1930, he began to visit to Princeton, initially as a visiting lecturer and then, starting from 1931, as a visiting teacher. In 1933, he accepted a permanent position there in the newly opened Institute for Advanced Study. Likely, the most important reason for von Neumann's choice of leaving Germany, aside from the natural antipathy for the Nazi regime, was the prospect that Princeton would replace Göttingen as the world center of research, where each scientist was free to deal with any topic of interest without any constraint.[4] Starting from 1932, the Institute tried to recruit the best mathematicians and physicists they could find, and attracted those Jewish and non-Jewish scientists who wanted to escape from a Europe on the brink of catastrophe. Albert Einstein and Hermann Weyl were among the first scientists to move to Princeton, and Weyl insisted that the Institute also offered von Neumann a position. Subsequently, a number of great scientists, among whom Paul Dirac, Wolfgang Pauli, and Kurt Gödel, joined the Institute, which became, as promised, a research center of worldwide importance.

From 1933, von Neumann spent most of his time in the United States. He became an American citizen in 1937 and started calling himself Johnny. The American period is full of important works that have touched different fields of research. It is not our purpose to cite all the important contributions of those years, but his works on functional analysis, in particular, those on the rings of operators that led to the introduction of what will be called von Neumann's Algebras, certainly deserve to be mentioned. Another important field of research that benefit of von Neumann's foundation contributions is the ergodic theory (see the entry Statistical Mechanics), respect to which von Neumann, with few works that exploited the theory of operators and functional analysis, managed to trace a promising way forward.

A turning point in von Neumann's life was the outbreak of World War II and the participation in the Manhattan project. Already before the war, after obtaining American citizenship, von Neumann began to collaborate with U.S. government in numerous civil and military institutions and during the World War II his involvement in military research increased, and he could barely find time for his free research activity. From September 1943, he was involved, as a hydrodynamics expert, in

---

[4]The mission of the Institute was clear from its presentation: *"The Institute for Advanced Study is one of the few institutions in the world where the pursuit of knowledge for its own sake is the ultimate* raison d'être. *Speculative research, the kind that is fundamental to the advancement of human understanding of the world of nature and of humanity, is not a product that can be made to order. Rather, like artistic creativity, it benefits from a special environment"*.

the Manhattan project, a secret project for the construction of the atomic bomb, directed by the physicist Robert Oppenheimer. The most brilliant scientists of the time were co-opted into the project and among others were present Niels Bohr, Enrico Fermi, Richard Feynman, Stanislaw Ulam, Leo Szilárd, Edward Teller, and Eugene Wigner. In the Manhattan project, the main activity of von Neumann was the theoretical development and the practical computation in fluid dynamics with particular reference to the dynamics of shock waves caused by explosions. It was a von Neumann's original idea the use of the implosion of nuclear fissile material as the atomic bomb ignition technique. For that purpose von Neumann's invented the explosive lenses, i.e., the use of a combination of both slow and fast explosives in order to focus the shock waves onto a spherical shape. Although mainly taken from practical activity von Neumann had not lost his touch and in his report on hydrodynamics, Theory of Shock Waves (1943), besides his own original result, he systematized in an organic and clear form all the ideas about the argument which until then were scattered in various papers some of which poorly accurate and with errors. Moreover, some other original contributions of von Neumann to fluid dynamics included the discovery of the classic flow solution to blast waves, and the derivation of the ZND detonation model of explosives (discovered independently by Yakov Borisovich Zel'dovich and Werner Döring).

Within the Manhattan project, his ideas which most profoundly impacted the development of technology in the following decades concern the structure of computers. Actually, in order to perform the complex numerical computations needed for the Manhattan project, the use of computers became fundamental. At first, the operators worked with desktop calculators, but from the end of 1943 an IBM computer with punched cards started to be used. Although the technology was very primitive, the calculations were slow and sometimes errors could occur and manual checks were necessary. von Neumann immediately understood the computer's potentiality and began to work both on numerical method and on the structure of computer.[5] The invention of von Neumann of the "stored-program computer", presented in his work "First Draft of a Report on the EDVAC" (1945) was a milestone in the computer development and shaped modern computers, for more details see the entry Computer, Algorithms, and Simulations.

Concerning numerical methods, the paper Numerical Inverting of Matrices of High Order (1947) in collaboration with Herman Goldstine can be considered as the beginning of modern numerical linear algebra, and it was the first paper addressing

---

[5] *"von Neumann played a leading role in the launching of electronic computers. His unique combination of gifts, his interests, and traits of character suited him for that role. I am thinking of his ability, and inclination to go through all the tedious details of program planning, of executing the minutiae of putting very large problems in a form treatable by a computer. It was his feeling for and knowledge of the details of mathematical logic systems and the theoretical structure of formal systems that enabled him to conceive of flexible programming. This was his great achievement. By suitable flow diagramming and programming, an enormous variety of problems became calculable on one machine with all connections fixed. Before his invention one had to pull out wires and reconnect plug boards each time a problem was changed"*. Stanislaw Ulam—Adventures of a Mathematician.

both the problem of rounding error and the philosophy of modern scientific computing. Another milestone in the applied science and computer simulations was given by
Princeton's *Meteorological Project*, started in 1946 and led by von Neumann. Such a
project, which began to realize Richardson's dream for the weather forecasting (see
the entry Richardson), for the first time made it possible to carry out daily numerical
weather predictions in less than 2 hour of ENIAC machine time. Together with the
introduction of specific computational algorithms, based on finite difference method
for the approximation of differential equations, von Neumann's main contribution
was the introduction of effective equations. About this last issue von Neumann and
his coworker Charney, a brilliant geophysicist, noted that the set of equations originally proposed by Richardson for the study of weather prediction contained too
much details, including meteorologically insignificant high-frequency component
(e.g., gravity waves), and therefore it was necessary to introduce an effective equation which filtered out such irrelevant variables. Separating the meteorologically
significant part of the phenomenon from the insignificant one brought an enormous
practical advantage: the numerical instabilities became less severe, hence relatively
large integration time steps could be used, at last making the numerical computations
satisfactorily efficient. In addition, even more importantly, the effective equations for
the slow dynamics capture the essence of the phenomena of interest, which could
otherwise be hidden in a too detailed description, as in the case of the complete
set of original equations. Moreover, he inspired the use of numerical simulation to
approach the difficult problem of turbulent flows, promoting computer simulations
as in silico experiments (see the entry Turbulence for further discussions).

   After the war, von Neumann's prestige increased further, as did both civil and mili-
tary government positions: director of the Electronic Computer Project at Princeton's
Institute for Advanced Study (1945–1955), President of American Mathematical
Society (1951–1952), from 1950 he became a consultant to the Weapons Systems
Evaluation Group (WSEG), from 1952 he became a consultant to the Central Intelligence Agency (CIA), and from 1955 he became a member of the Atomic Energy
Commission (AEC), just to mention some of his numerous assignments. He had also
been criticized for accepting such a high number of responsibilities as if he liked to
collect important positions, but his most powerful drive has always been to improve
the functioning of public systems, especially the military one given the inefficiency
and approximation that reigned in that area of capital importance especially after the
development of the atomic weapon. In this respect, it is interesting to contrast the
attitude of von Neumann with that of the pacifist Richardson (see the related entry)
as an example of the different attitudes of the human being behind the scientists.

   Unfortunately, the humankind could not benefit of von Neumann's abilities for a
long time because in 1955 he was diagnosed with cancer already in the metastatic
phase and died in February 1957. We conclude with the words of Stanislaw Ulam,
one of his dearest friends: *"In John von Neumann's death on February 8, 1957, the
world of mathematics lost a most original, penetrating, and versatile mind. Science
suffered the loss of a universal intellect and a unique interpreter of mathematics,
who could bring the latest (and develop latent) applications of its methods to bear
on problems of physics, astronomy, biology, and the new technology"*.

Chapter 31
A Random Walk in Fiction

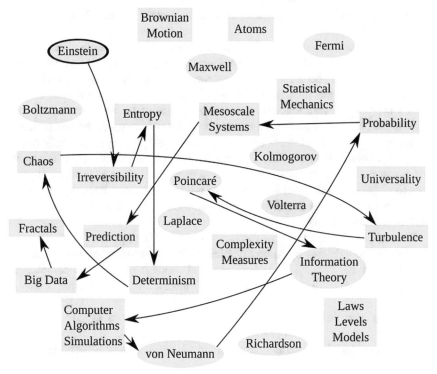

Here—with the mere aim of entertaining the reader and without any claim of completeness—we wish to illustrate a different itinerary through our book. This path is based on fictional works, e.g., movies and novels involving the concepts (and scientists) collected in this book. We invite the readers to search among the films they have seen or the books they have read for further connections with the physical theories and the authors presented in this book.

© Springer Nature Switzerland AG 2021
M. Cencini et al., *A Random Walk in Physics*,
https://doi.org/10.1007/978-3-030-72531-0_31

We begin from the boundary of our scientific domain, i.e., from Einstein. Einstein's character is so pop that has inspired several movies also somehow irreverent. For instance, in "Insignificant" (1985) directed by Nicolas Roeg, Einstein (Michael Emil) is fascinated by another pop figure, namely, Marilyn Monroe with whom he spends a night in a hotel discussing about universe and relativity in underwear. In I.Q. (1994) directed by Fred Schepisi, Einstein starred by an always funny Walter Matthau helps, with the aid of some colleagues (including an improbable Kurt Gödel), the mechanic Ed (Tim Robbins) to conquer the heart of Catherine (Meg Ryan), Einstein's nephew, who is a promising mathematician. She is engaged with a Psychologist (Stephen Fry), who, according to Einstein, does not love her enough. There are some hilarious moments in which Einstein and colleagues help Ed to appear to Catherine's eyes as an intellectual. Several science-fiction books and movies are based upon the widely known Einstein's results, a chief example is given by the recent blockbuster movie "Interstellar" by Christopher Nolan (2014), which heavily takes advantage of general relativity and the theory of black holes.

Interestingly, the last (and very debated) movie by the same director Christopher Nolan, "Tenet" (2020), is centered upon a fundamental concept which is discussed in our book: Irreversibility. The movie is quite controversial and, in a sense, defies the accepted idea that reversed trajectories are practically impossible to be observed in the macroscopic world. Amusingly, a concept strictly associated to irreversibility, which is entropy, has been involved in the romance movie "Entropy" (1999), where also the singer of the rock band U2, Bono, plays a role. In the movie, entropy evokes the popular idea that everything—in the life of the main character—gets worst and worst. Again a romance movie inspired by entropy and the second law of thermodynamics is "The laws of thermodynamics" (2018), where a physicist tries to apply such laws to his love story. The movie also presents other aspects of science and makes parallels between the relationships of human beings and several physical laws (e.g., gravitation). According to one of the characters, not knowing the second law of thermodynamics demonstrates the same lack of education as not knowing anything of Shakespeare.[1]

Romance is also the genre of the famous movie "Sliding doors" (1998), based on a movie about destiny by Krzysztof Kieślowski, "Przypadek" or "Blind Chance" (1981), where the concept of determinism and chaos is exploited to compare the two possible lives of the protagonist, evolving after a tiny discriminating event—a sort of Lorenz's butterfly effect. Moreover "The Butterfly effect" (2005) is the title of science-fiction thriller movie inspired to the same idea: a young guy suffers of a sort of blackout moments and, become adult, he discovers that he can go back to his youth in those blackout moments and changing a few small things of his past provokes

---

[1] This is likely inspired to *The Two Cultures* of Charles Percy Snow, English novelist and physical chemist, when he writes *"A good many times I have been present at gatherings of people who, by the standards of the traditional culture, are thought highly educated and who have with considerable gusto been expressing their incredulity at the illiteracy of scientists. Once or twice I have been provoked and have asked the company how many of them could describe the Second Law of Thermodynamics. The response was cold: it was also negative. Yet I was asking something which is the scientific equivalent of: Have you read a work of Shakespeare's?"*.

several alternatives to his future. The action movie "Chaos" (2005) makes explicit reference both to Lorenz, which is the name of the main character (Wesley Snipes) although being a criminal in the film, and to the bestseller of James Gleick *Chaos: Making a New Science*. Here chaos is used as an argument of conversation between the criminals and the detective Conners (Jason Statham) who negotiates with them for some hostages. An attempt to involve chaos in a more precise way appears in the novel "Jurassic Park" by Michael Crichton (1990) and in the blockbuster movie based on it directed by Steven Spielberg (1993), where one of the protagonists is the mathematician and chaos expert Ian Malcolm. The point of view of this character is that instabilities due to chaos make very hard to contain and control the bizarre experiments of the creator of the park, leading to unexpected and dangerous results. Another blockbuster movie that exploits physical concepts discussed in this book is "Turbulence" (1997), where the physics of strong randomness enters again as a dramatic source of disasters and, of course, special effects. Turbulence, or, more precisely, the Navier-Stokes equation, has been a source of inspiration for "Gifted" (2017) about the life of a math gifted baby-girl, whose mother was a brilliant mathematician who dedicated her life to solving the millennium problem related to the Navier-Stokes equation[2] (with a prize of 1 million dollars). The mother fails to solve the problem and commits suicide, but it is not the end of the story and the movie has an unexpected ending.

Subtle and interesting is the use of Poincaré's results on the three-body problem in the science-fiction book "The Three-Body Problem" by the Chinese writer Liu Cixin (2008). The plot involves an alien population living in a distant system with three suns, therefore obliged to survive cyclical (non-periodical) devastations. "The Three-Body Problem: a Cambridge mystery" is also the title of a book by Catherine Shaw (the pseudonym of an unknown mathematician) (2004) who exploits the idea of a thriller story about a mysterious triple murders—three corpses in the story. However, the main characters is in love with mathematics and in the novel she meets historical mathematicians, including Arthur Cayley and Gösta Mittag-Leffler (see the entry Poincaré) and who have to solve the mystery linked to the n-body problem and to the social aspects of mathematical collaborations.

Information, which in theoretical physics is technically related to entropy, has a major role in several books and movies telling the fascinating role of cryptography in the World War II. Among the most famous we recall "Enigma", which is a novel by Robert Harris (1995) translated in a movie (2001), followed more recently by other blockbusters such as "Codebreaker" (2011) and "The Imitation Game" (2014) specifically on the life of Alan Turing who, besides his role as code breaker of the Enigma machine, have had a key role in theoretical computer science and on the birth of electronic computers. Computers and "human" calculators are the subject of "Hidden Figures" (2016) directed by Theodore Melfi, which tells the story of the human calculators (three black women) in assisting the NASA Space Task Group which was under pressure after Soviet Union's successful launch of Yuri Gagarin

---

[2]"The Navier-Stokes existence and smoothness problem" concerns the proof of the properties, existence, and smoothness, of the solutions of the Navier-Stokes equation.

into space. Besides narrating the hidden role of these women and the transition from human to automatic calculators (one of the protagonists Dorothy Vaughan learns Fortran to use IBM 7090 bought from NASA), the movie is interesting for the way gender and racial issues are analyzed. Above all, the movie is based on the true life of Katherine Johnson who was awarded the Presidential Medal of Freedom (2015) for her role in NASA space missions. A beautiful and philosophical movie on the role of the computer for forecasts is Krzysztof Kieślowski's "Dekalog One" (1988), connected to the first imperatives *"I am the Lord your God; you shall have no other gods before me"*. The protagonist, Krzysztof, is a physicist, university professor, and a great computer enthusiast. He believes that physical world can be described mathematically through a computer code, but blind faith in the results of a numerical simulation can have devastating consequences.

"A beautiful mind" (2001) about the story of John Nash, a mathematician only mentioned in this book, is set initially in Princeton, in the same period in which Albert Einstein and John von Neumann were also present. John Nash perfected von Neumann's game theory, and such a topic is also present in "Dr. Strangelove or: How I Learned to Stop Worrying and Love the Bomb" by Stanley Kubrick (1964). In this wonderful movie, a ferocious satire of the cold war, likely the main character is a sort of mockery of von Neumann. Actually, in the movie, there is a direct mention of mutual assured destruction (MAD), which is a theory developed by John von Neumann about the equilibrium strategy (a Nash's equilibrium) which is achieved when two conflicting powers have at their disposal an atomic arsenal sufficient to end life on earth.

Probability has been a source of inspiration for "21" (2008) an American heist drama film directed by Robert Luketic based on Ben Mezrich's book *Bringing Down the House,* the best-selling book about the true story of a group of MIT students who, in team, exploit their abilities with counting and probabilistic reasoning to increase the probability to win at blackjack in Las Vegas casinos. The protagonist is a brilliant student, aiming to move to Harvard Medical School, who is noted by the math teacher (the head of the group, starred by Kevin Spacey) when he gives instantly the proper answer to the Monty Hall problem.[3] Of course, watching the film, likely, the reader would be tempted to study probability to become rich by playing blackjack, we suggest to resist to this temptation but not to that of studying probability, whose usefulness goes much beyond games of chances. Yet on probability with applications of Bayes principle and, possibly, some tools mentioned in the entry Big Data, is the movie "Moneyball" (2011) directed by Bennett Miller and with Brad Pitt, based on Michael Lewis' book with the same title. It narrates the (true) story of Billy Beane

---

[3]The Monty Hall problem is quite a famous probability puzzle, used in some TV shows, which works as follows. There are three doors, one leads you to a brand new car the other two to two goats. You are asked to choose a door. Then the game conductor, who knows where the goats are, open among the two remaining doors the one with the goat behind. Finally, you are given the possibility either to maintain your choice or to choose the other remaining door. An incorrect probabilistic reasoning would suggest that you have 50% chances to have chosen the correct door, while with the correct reasoning you can realize that you have two over three chances to find the car by changing the door. We invite the reader to find out why.

which uses statistics and probabilistic reasoning to assemble a competitive team for the Oakland Athletics baseball resorting to many players which were overlooked because of age, personality, or other reasons.

Mesoscale systems, that is, physical systems at a scale which is not astronomical but neither atomic, are brilliantly treated in the "Fantastic Voyage" movie (1966) and the subsequent novel by Isaac Asimov (same year with the same title).

We conclude this non-exhaustive overview, mentioning the beautiful "Minority report" short story by Philip K. Dick (1956) translated—by Steven Spielberg—in the movie with the same title in 2002. The story tells of some mutant characters ("precogs") which are able to predict crimes, therefore intimately related to our entry Prediction. It is however amazing to realize that Dick describes the role of these *precogs* as cryptic oracles which produce data (on future crimes or other facts) which are difficult to be unraveled and can also be conflicting one with the others, requiring human assistance—and some arbitrariness—for their interpretation. In our opinion such a scenario is not far from some perspective illustrated in our entry Big Data. Yet about prediction there is "The Bank" (2001) by Robert Connolly, here the idea is that the mathematics of fractals and in general of complex systems can be used to predict the stock market. During the film credit images taken from famous fractal sets (the Mandelbrot set) are shown with an evocative music in background.

Finally, as far as literature is concerned, many great writers have drawn heavily from scientific culture, from Borges to Calvino, from Queneau to McEwan, up to the postmodernist Pynchon and Foster Wallace, but unfortunately this discussion would take us too far from the aims of this book which has come to an end.

Further Reading

Great Scientists

O.M. Ashford, *Prophet or Professor? The Life and Work of Lewis Fry Richardson* (Adam Hilger, Bristol, 1985)

C. Bartocci, R. Betti, A. Guerraggio, R. Lucchetti (eds.), *Mathematical Lives: Protagonists of the Twentieth Century from Hilbert to Wiles* (Springer, Berlin, 2011)

L. Campbell, W. Garnett, *The Life of James Clerk Maxwell* (MacMillan and Co., London, 1882)

C. Cercignani, *Ludwig Boltzmann: the Man Who Trusted Atoms* (Oxford University Press, Oxford, 1998)

E. Charpentier, A. Lesne, N.K. Nikolski (eds.), *Kolmogorov's Heritage in Mathematics* (Springer, Berlin, 2004)

D. Cooper, *Enrico Fermi* (Oxford University Press, Oxford, 1998)

L. Fermi, *Atoms in the Family: My Life with Enrico Fermi* (University of Chicago Press, Chicago, 1954)

P. Galison, *Einstein's Clocks and Poincaré' s Maps: Empires of Time* (W.W. Norton & Company, New York, 2004)

S. Gindikin, *Tales of Mathematicians and Physicists* (Springer, Berlin, 2007) [Among other also Laplace, Lagrange and Poincaré are described]

N.P. Gleditsch (ed.), *Lewis Fry Richardson: His Intellectual Legacy and Influence in the Social Sciences* (Springer, Berlin, 2020)

A. Guerraggio, G. Paoloni, *Vito Volterra* (Springer Italia, 2013)

G. Israel, A. Millan Gasca, *The World as a Mathematical Game: John von Neumann and Twentieth Century Science* (Birkhäuser, Basel, 2009)

D. Lindley, *Boltzmann's Atom* (Free Press, New York, 2001)

R. Livi, A. Vulpiani (eds.), *The Kolmogorov Legacy in Physics: A Century of Turbulence and Complexity* (Springer, Berlin, 2003)

B. Mahon, *The Man Who Changed Everything* (Wiley, Chichester, 2003) [This book is about J.C. Maxwell]

J. Mehra, *The Golden Age of Theoretical Physics* (World Scientific, Singapore, 2001)

© Springer Nature Switzerland AG 2021
M. Cencini et al., *A Random Walk in Physics*,
https://doi.org/10.1007/978-3-030-72531-0

A. Pais, *Subtle is the Lord... The Science and the Life of Albert Einstein* (Oxford University Press, Oxford, 1982)

F. Verhulst, *Henri Poincaré Impatient Genius* (Springer, Berlin, 2012)

A. Vulpiani, Lewis Fry Richardson: scientist, visionary and pacifist. Lett. Mat. Int. Ed. **2**, 121 (2014)

Atoms and Matter

S.G. Brush, A history of random processes: Brownian movement from Brown to Perrin. Arch. Hist. Exact Sci. **5**, 1 (1968)

L. Cohen, The history of noise. IEEE Signal Process. Mag. **20** (2005)

M. Haw, *Middle World: the Restless Heart of Matter and Life* (Macmillan, London, 2007)

R. Jones, *Soft Machines* (Oxford University Press, Oxford, 2004)

M.J. Nye, *Molecular Reality* (Macdonald, London, 1972)

J. Perrin, *Les Atomes* (Alcan, Paris, 1913); [English translation] *Atoms* (Van Nostrand, New York, 1916)

R. Piazza, *Soft Matter: the Stuff That Dreams Are Made Of* (Springer Italia, 2010)

Prediction, Models and Laws

F. Cecconi, M. Cencini, M. Falcioni, A. Vulpiani, The prediction of future from the past: an old problem from a modern perspective. Am. J. Phys. **80**, 1001 (2012)

M. Davis, *The Universal Computer: the Road from Leibniz to Turing* (Norton and Company, New York, 2000)

H. Goldstine, *The Computer. From Pascal to von Neumann* (Princeton University Press, Princeton, 1993)

P. Lynch, *The Emergence of Numerical Weather Prediction: Richardson's Dream* (Cambridge University Press, Cambridge, 2006)

A. Rényi, *Dialogues on Mathematics* (Holden-Day, San Francisco, 1967)

L.F. Richardson, *Weather Prediction by Numerical Methods* (Cambridge University Press, Cambridge, 1922)

L.J. Snyder, *The Philosophical Breakfast Club* (Broadway Books, New York, 2011) [In this book there is a nice discussion of the tidal prediction in the victorian period]

Chaos, Fractals, Information and Complexity

D. Aubin, A.D. Dalmedico, Writing the history of dynamical systems and chaos: Longue Durée and revolution, disciplines and cultures. Hist. Math. **29**, 273 (2002)

J. Bricmont, Science of chaos or chaos in science? *Flight from Science and Reason.* Ann. N. Y. Acad. Sci. **775**, 131 (1996)

J. Ford, How random is a coin toss? Phys. Today **36**, 40 (1983)

A. Gabrielli, F. Sylos Labini, M. Joyce, L. Pietronero, *Statistical Physics for Cosmic Structures* (Springer, Berlin, 2005)

K. Gleick, *Chaos: Making a New Science* (Viking Penguin, New York, 1987)

K. Gleick, *The Information: a History, a Theory, a Flood* (Pantheon Books, New York, 2011)

P. Grassberger, Toward a quantitative theory of self-generated complexity. Int. J. Theor. Phys. **25**, 907 (1986)

E.N. Lorenz, *The Essence of Chaos* (UCL Press, London, 1993)

B. Mandelbrot, *Fractals: Form, Chance, and Dimension* (Freeman, New York, 1977)

D. Ruelle, *Chance and Chaos* (Princeton University Press, Princeton, 1991)

Probability

P. Diaconis, B. Skyrms, *Ten Great Ideas About Chance* (Princeton University Press, Princeton, 2018)

B.V. Gnedenko, A.Ya. Khinchin, *An Elementary Introduction to the Theory of Probability* (W.H. Freeman and Company, New York, 1961)

M.P. Silverman, *A Certain Uncertainty: Nature's Random Ways* (Cambridge University Press, Cambridge, 2014)

J. Tabak, *Probability and Statistics* (Facts On File, Inc., 2004)

R. Von Mises, *Probability, Statistics and Truth* (Macmillan, New York, 1957)

Big Data

C. Anderson, The end of theory: the data deluge makes the scientific method obsolete. Wired (2008), http://www.wired.com/2008/06/pb-theory/

M. Buchanan, The limits of machine prediction. Nat. Phys. **15**, 304 (2019)

P. Domingos, *The Master Algorithm* (Basic Books, New York, 2015)

T. Hey, S. Tansley, K. Tolle (eds.), *The Fourth Paradigm: Data Intensive Scientific Discovery* (Microsoft Research, Redmond, 2009)

H. Hosni, A. Vulpiani, Forecasting in light of big data. Philos. Technol. **31**, 557 (2018)

M. Mezard, Artificial intelligence and its limits. Europhys. News **49**, 26 (2018)

Statistical Mechanics, Turbulence and Universality

P.W. Anderson, More is different. Science **177**, 393 (1972)

R. Batterman, *The Devil in the Details: Asymptotic Reasoning in Explanation, Reduction, and Emergence* (Oxford University Press, Oxford, 2002)

M.V. Berry, Asymptotics, singularities and the reduction of theories. Log. Methodol. Philos. Sci. **IX**, 597 (1994)

S.G. Brush, *Statistical Physics and the Atomic Theory of Matter from Boyle and Newton to Landau and Onsager* (Princeton University Press, Princeton, 1983)

S. Chibbaro, L. Rondoni, A. Vulpiani, *Reductionism, Emergence and Levels of Reality* (Springer, Berlin, 2014)

U. Frisch, *Turbulence: the legacy of AN Kolmogorov* (Cambridge University Press, Cambridge, 1995)

M. Eckert, *The Turbulence Problem* (Springer, Berlin, 2019)

G. Gallavotti, *Statistical Mechanics. A Short Treatise* (Springer, Berlin, 1995)

J.L. Lebowitz, Boltzmann's entropy and time's arrow. Phys. Today **46**, 32 (1993)

S.-K. Ma, *Statistical Mechanics* (World Scientific, Singapore, 1985)

I. Marusic, S. Broomhall, Leonardo da Vinci and fluid mechanics. Annu. Rev. Fluid Mech. **53**, 1 (2021)

S. Weinberg, Newtonianism, reductionism and the art of congressional testimony. Nature **330**, 433 (1987)

K.G. Wilson, Problems in physics with many scales of length. Sci. Am. **241**, 140 (1979)

Printed in the United States
by Baker & Taylor Publisher Services